Springer Textbooks in Earth Sciences, Geography and Environment

The Springer Textbooks series publishes a broad portfolio of textbooks on Earth Sciences, Geography and Environmental Science. Springer textbooks provide comprehensive introductions as well as in-depth knowledge for advanced studies. A clear, reader-friendly layout and features such as end-of-chapter summaries, work examples, exercises, and glossaries help the reader to access the subject. Springer textbooks are essential for students, researchers and applied scientists.

More information about this series at http://www.springer.com/series/15201

Paul Alexandre

Isotopes and the Natural Environment

 Springer

Paul Alexandre
Geology Department
Brandon University
Brandon, MB, Canada

The Instructors' Solution Manual can be downloaded from https://springer.com/9783030336516.

ISSN 2510-1307   ISSN 2510-1315   (electronic)
Springer Textbooks in Earth Sciences, Geography and Environment
ISBN 978-3-030-33654-7   ISBN 978-3-030-33652-3   (eBook)
https://doi.org/10.1007/978-3-030-33652-3

This Springer imprint is published by the registered company Springer Nature Switzerland AG
The registered company address is: Gewerbestrasse 11, 6330 Cham, Switzerland

To my wife, for all she's taught me.

Preface

This book is partially derived from—and based upon—a course by the same name that I used to deliver at Queen's University in Canada. It was a course that I loved teaching, but it was not always very straightforward for students: they found it at first intimidating (Isotopes? What's that?), then complex, then downright difficult. However, by the end of the course each and every student learned to appreciate the significance and importance of isotopes when used as an investigative tool in the natural sciences.

To help students recognise the importance of isotopes in the natural sciences, I devised a simple game, in the form of a bet. At the very beginning of the semester, during my introductory lecture, I will explain that isotopes are applicable to absolutely every branch of the natural sciences. Then I will offer them the bet (or was it a bait?): If a student manages, by the end of the semester, to find an example, any example, of a field belonging to the natural sciences where isotopes are not of any use, have never been of any use, and likely will never be of any use, that student will receive a 100% mark for the course; this is upon condition that the example is well substantiated.

Students invariably accepted the bet: they were certain that there must be something out there that is not, in any way, related to isotopes. Just as invariably, no student ever claimed the prize. When I reminded them of our bet, during the last lecture, and invited them to step forward and attempt it, they ruefully replied that there is no way they can win (and sometimes even hinted that the bet was not fair). As they put it, well, everything material is made of isotopes and these isotopes constantly change (through fractionation or disintegration), meaning that every process that has affected a material object will be reflected in the isotopic composition of that object.

This is, in a nutshell, the topic and the message of this book. Isotopes are in absolutely any material object, from a very simple life-form to the entirety of planet Earth and, of course, beyond. Think of it for a second: each time you breathe (including right now as you read these words), oxygen isotopes from the air fractionate (change isotopic composition) in your lungs. When you are in the street and it rains or snows, hydrogen and oxygen isotopes in the water fractionate. Importantly, the extent of fractionation and the factors affecting it are well known and understood: if we measure the isotopic composition of an object (the air, the rain water), we can have a very clear idea of the processes that affected it and, significantly, the physical and chemical conditions at which these processes occurred.

The book begins by explaining what isotopes are and how they are affected by any sort of processes and by the passage of time. We will then briefly visit the laboratory and have a quick look at how isotopic compositions are measured and reported. After this, let the fun begin: we will visit plenty of examples, from a variety of natural sciences fields, of applications of isotopes as an investigative tool. These will vary from the geosphere (geology, or the rock realm), to the hydrosphere, the atmosphere (with a word or two about climate change), to the biosphere (animals and plants), and to human health and human evolution and history.

However, allow me a word of caution: it is absolutely impossible to include examples of every possible use of isotopes as an investigative tool in this introductory and general book. Such a task would run into thousands of pages and would be, ultimately, unreadable. Let us rather settle on a lighter approach, picking things up here and there as we go, with a few simple and straightforward examples, making everything easy to understand and helping you appreciate the big picture, namely, do not try that bet!

A word of recognition: I have been momentously helped on the way by countless people. Credit goes to my family, who have been very patient and helpful; to my colleagues, for all their support, and to the Springer editorial board, for all the practical advice and help, but also for their support and understanding. To all of these, thank you!

Brandon, Canada   Paul Alexandre
July 2019

Contents

Part I
Theory and Methodology of Isotopes

1.1 Nature and Formation of Isotopes

What are Isotopes?

The very first thing that we have to examine is the nature of isotopes: what they are, when and how they were formed, and what their characteristics are. The word isotope itself means "same place", from the Greek words ισοσ ("same") and τοποσ ("place"), and refers to their position in the periodic table of the elements. Each element is defined by its atomic number, going from 1 (hydrogen, H) to 109 (meitnerium). We will not consider the extremely short-lived and mostly synthetic elements above atomic number 109. This number corresponds to the number of protons in the atomic nucleus. In a neutral, non-ionized atom, the number of protons (charge of +1) is equal to the number of electrons (charge of −1). Given that the chemical characteristics and behaviour of the elements are—to a very large extent—defined by their number of electrons (and their electronic orbital configuration), it follows that the chemical behaviour of different elements will be conditioned and defined by their number of protons.

Another important particle is present in the nucleus of an atom: the neutron. For our purposes, neutron's most important characteristics are its mass of 1 and its electroneutrality (it has no charge). As a result, adding additional neutrons to a nucleus of an atom will not modify its chemical characteristics and behaviour in the natural environment. Thus, *isotopes are variants of the same chemical element differing uniquely by the number of neutrons in their atomic nucleus.*

In other words, the different isotopes of the same chemical element will be found in the exact same position in the periodic table, meaning that they will have the same chemical characteristics, such as electronic configuration, ionic radius, valence states, electronegativity, first ionization potential, and so on. These characteristics will, in turn, condition the chemical behaviour of the elements. Impor-

tantly, the different isotopes of the same element will have the exact same chemical behaviour. It is important to repeat this and to keep it in mind at any point: the different isotopes of the same element differ only by the number of neutrons in their nucleus—and thus by their mass—but have the same chemical behaviour (Fig. 1.1).

The mass of each individual isotope is given by the sum of the number of protons and the number of neutrons present in its atomic nucleus. For our purposes, mass is the most important and defining characteristic of isotopes. It is the mass that will affect the behaviour, in the natural environment, of the different isotopes of the same chemical element that otherwise will have the same chemical characteristics.

The official name of the unit of weight used for isotopes is Dalton (abbreviated to Da, or u), after John Dalton, the famed chemist and mineralogist at the origin of modern atomic theory. The weight of 1 Da is defined as one-twelfth of the mass of the unbound ^{12}C at rest and in ground state and is equal to $1.6605390402 \times 10^{-24}$ g. More commonly, isotope scientists speak of Atomic Mass Unit (AMU) as identical in weight, or they choose not to mention the unit at all, something which we will often do in this book (Fig. 1.2).

Several, approximately a quarter of all chemical elements, have only one isotope; we call these mono-isotopic elements (Be, F, Na, Al, P, Sc, Mn, Co, As, Y, Nb, Tc, Rh, I, Cs, Pr, Pm, Tb, Ho, Tm, Au, Bi, Th, and Pa). They are, in most circumstances, not useful for any natural environment-related study using isotopes, and will not be considered further, with exception to the radiogenic ones (e.g., ^{232}Th, which ultimately disintegrates to the stable ^{208}Pb). Many other isotopes are very short lived, often milliseconds to a few seconds, and this makes them unsuitable for natural environment-related studies, similarly to those synthesized in laboratory conditions; these also will not be considered here. The full list of naturally occurring isotopes, along with their exact masses and proportions, is given in Appendix A (Fig. 1.3).

© Springer Nature Switzerland AG 2020
P. Alexandre, *Isotopes and the Natural Environment*, Springer Textbooks in Earth Sciences, Geography and Environment, https://doi.org/10.1007/978-3-030-33652-3_1

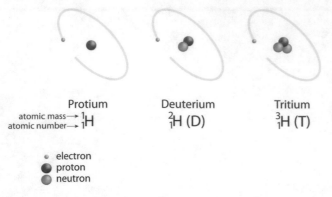

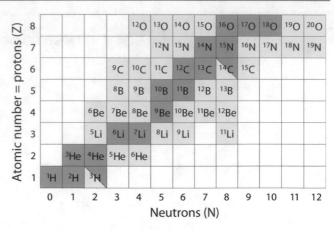

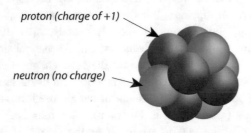

Fig. 1.1 The three main isotopes of hydrogen, the simplest of elements: protium (or just hydrogen, ^{1}H), made of one proton and one electron, deuterium, made of one proton, one neutron, and one electron (^{2}H), and the radioactive (or unstable) tritium, made of one proton, two neutrons, and one electron (^{3}H). The common factor in each case is the number of protons, equal to the number of electrons, giving the atomic number of 1; the difference is the number of neutrons, from 0 for ^{1}H to 2 to ^{3}H. Thus, the three isotopes will have the exact same chemical characteristics and behaviour (defined by the number of protons), but different masses

Fig. 1.3 The first eight elements in the chart of isotopes, where the nuclei are plotted by their number of protons and number of neutrons: each line corresponds to one *chemical element* and its various isotopes, each occupying one box and defined by their mass. Blue boxes denote stable isotopes and yellow boxes mean radioactive isotopes. Two isotopes are considered stable but are in fact radioactive: ^{3}H and ^{14}C are produced (by cosmic radiation in the upper atmosphere) at approximately the same rate at which they disintegrate, keeping their amount on Earth approximately constant

Fig. 1.2 Schematic representation of the nucleus of ^{11}B (pronounced *boron 11*), containing 5 protons and 6 neutrons, for atomic number (Z) of 5 and atomic mass 11

prevent it from disintegration. We can see this concept illustrated in Fig. 1.4, where the line of stability moves away from the N = Z line. Maximum stability is achieved when a specific number of protons and neutrons are present in the nucleus. Isotopes with a different combination of protons and neutrons will stray away from the line of stability and will be unstable, tending to disintegrate: the further they are from the line of stability, faster they will disintegrate.

Line of stability

The neutrons fulfil a very important function, which is to hold the nucleus together. The particles making up the nucleus, protons, and neutrons, are subjected to two opposite forces: the electrostatic force and the so-called strong interaction. The former affects only the protons and is repulsive (pushing two protons away from each other) while the latter works between protons and neutrons and is attractive (attracting neutrons and protons towards each other). It is the combination of these two forces that determines how stable a nucleus is. Importantly, the two forces do not act at the same distance: the electrostatic force is longer range than the strong interaction. Thus, with heavier isotopes, we must add even more neutrons to the nucleus than protons to keep the nucleus stable and

A Little Bit of Nomenclature

Generally, isotopes are placed in one or more of the following groups based on their mass, their stability, and their origin; an isotope can belong to more than one category:

- Light isotopes: isotopes of elements with atomic number below 8 (H, He, Li, Be, B, C, N, and O). These are very abundant on Earth's surface, in the hydrosphere, and in the biosphere; they form the bulk of matter in the universe and are extensively studied. They are stable, with the exception of ^{3}H and ^{14}C.
- Heavy isotopes: isotopes of elements with atomic number typically above 70 (Hf, Ta, W, Re, Os, Ir, Pt, Hg, Tl, Pb, Th, and U), with several being radioactive (unstable).

Fig. 1.4 The line of stability is defined by the most "advantageous" combination of the number of protons and the number of neutrons that will produce the most stable nuclei. As more protons are added to make heavier elements, even more neutrons are required to counteract the repulsion by the longer range electrostatic force: in this case, the line of stability will go further away from the N = Z line

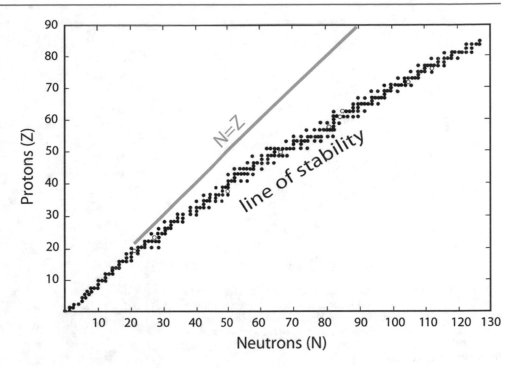

They are predominantly found in the geosphere (the solid Earth).

- Stable isotopes: the number of protons and neutrons in these nuclei will not be modified under any naturally occurring circumstances and processes on Earth: their nucleus will remain strictly as it is.
- Radioactive isotopes: they have unstable nuclei and disintegrate at a constant rate by emitting particles and energy.
- Radiogenic isotopes: produced by the disintegration of radioactive isotopes. A radiogenic isotope can itself be unstable and further disintegrate, or be stable (in which case it will remain as it is).
- Cosmogenic isotopes: produced by cosmic radiation in the upper reaches of the atmosphere. The most famous example is ^{14}C, which is also radioactive, making it useful in "carbon dating", which we will consider later.

Nucleosynthesis
Simplistically, there are four main settings where nuclides were and are produced:

1. During the Big Bang, between one second and three minutes after the beginning of the universe, when temperatures were sufficiently low (around 10^9 K) to allow the formation of nuclear particles and complex nuclei. Firstly, all of ^1H (consisting of one proton) was formed at that point, followed by ^2H and ^3H (addition of one and two neutrons,

respectively), ^3He and ^4He, and minute amounts of Li, Be, and possibly B. Temperature and, to a lesser extent, pressure were too high for any other nuclide to be stable.

2. Stellar nucleosynthesis, which occurred in a star from the main sequence and consisted of three episodes of nuclear fusion, whereby nuclides combined to form heavier and more stable isotopes, releasing energy in the process.

   - Episode 1, proton addition. Small amounts of ^2H and ^3H were produced, together with ^3He and ^4He, by the major and very slow proton–proton reaction (PPI). This was followed by the minor PPII and PPIII processes, responsible for the formation of isotopes up to ^{13}C. This stage lasts millions of years.
   - Episode 2, the C–N–O cycle. A series of proton additions, combined with γ and β^+ decays, formed all C, O, and N isotopes. This stage is faster: it lasts thousands to hundreds of years.
   - Episode 3, carbon, oxygen, and nitrogen burning. By using C, O, and N as the main "fuel" and combining them with lighter elements or individual particles, all isotopes up to ^{56}Fe were formed. Things are much faster now: these stages last years to days.

A major characteristic of stellar nucleosynthesis is that it occurs in conditions of mechanical equilibrium, resulting in the formation of predominantly stable

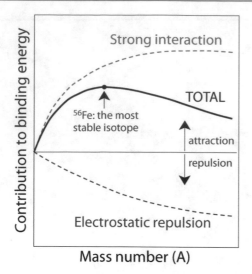

Fig. 1.5 The total binding energy for all elements, resulting from the combination between the electrostatic force and the strong interaction, plotted against atomic number. The highest total binding energy is observed for ⁵⁶Fe, which is the most stable isotope. As the effects of the electrostatic repulsion between protons exceed the attractive strong interaction, nuclei become less stable

isotopes: this is why the light isotopes are predominantly stable, with very few exceptions. The most stable isotope, ⁵⁶Fe, is also the last to be produced in a star from the main sequence. If masses heavier than 56 are formed, they are unstable and disintegrate to ⁵⁶Fe (Fig. 1.5).

Another characteristic of stellar nucleosynthesis is the constant release of energy due to nuclear fusion and isotopes occupying lower energy states. A large proportion of this energy will be emitted outside of the star, but some will cumulate, leading eventually to a brutal departure from the state of equilibrium and the formation of Super Nova.

3. Super Nova: explosive nucleosynthesis. We are no longer in equilibrium, and a rapid addition of neutrons (the s-process, occurring in a semblance of equilibrium) is combined with β⁻ decay (the r-process, in disequilibrium). These two processes form all the remaining naturally occurring isotopes, all the way to ²³⁸U. Significantly, the s-process will tend to form stable isotopes (the heaviest of which is ²⁰⁹Bi), while the r-process will create mostly radioactive isotopes.

Here we have the main reason for the existence of stable and radioactive isotopes in that the stable isotopes form under conditions of mechanical equilibrium (in the star) or under conditions approaching equilibrium (the s-process). Radioactive isotopes are formed under conditions of disequilibrium, not appropriate for the formation of nuclei containing the best numbers of protons and neutrons allowing them to be stable.

There is another setting where elements are formed: outer space, where spallation occurs. The process consists of nuclei colliding with a high-energy cosmic ray (consisting mainly of fast protons) and producing small amounts of light elements, such as ³H, ⁶Li, ⁹Be, ¹⁰B, and ¹¹B. These isotopes are actually consumed at temperatures involving hydrogen burning in a star and are unstable at Big Bang conditions, which is why spallation is the only process that produces them. However, with collision probability in outer space being exceedingly low, we end up with very low amounts of these isotopes.

Isotopic Composition

By convention, the isotopic composition of a sample is expressed relative to that of a specific standard, which is different for different elements. In all cases, we are interested in the ratio of the minor, heavier isotope versus the lighter, major one, for instance $^2H/^1H$, or $^{18}O/^{16}O$, or $^{13}C/^{12}C$, and so on. The ratios are compared to those of a standard, and are expressed by the δ-notation ("delta notation"), using the following equation, for the example of oxygen isotopes:

$$\delta^{18}O_{SAMPLE} = \left[\frac{\left(^{18}O/^{16}O\right)_{SAMPLE} - \left(^{18}O/^{16}O\right)_{STANDARD}}{\left(^{18}O/^{16}O\right)_{STANDARD}} \right] \times 10^3, ‰$$

The practical meaning of this notation is that a high δ value indicates a higher proportion of the heavier isotope relative to a standard. A positive δ value indicates that the sample is enriched in the heavy isotope relative to the standard, whereas a negative δ value indicates that the heavy isotope is depleted in the sample relative to the standard. When a sample has a δ value of 0 it indicates that the sample has the same proportion of the heavy isotope as the standard.

We must always indicate what standard we used to compare our sample. For the main light elements (H, O, C, N, and S)—elements extensively used in the natural environment studies—the standards are the following:

- for H and O: ocean water; specifically the Standard Mean Ocean Water (SMOW), with absolute ratios $^{2}H/^{1}H$ of 155.76×10^{-6} and $^{18}O/^{16}O$ of 2005.20×10^{-6};
- for C: the calcite of a specific fossil, belemnite, from the Peedee Formation in South Carolina, USA, abbreviated as PDB and having a ratio $^{13}C/^{12}C$ of 11237.2×10^{-6};
- for N: atmospheric nitrogen, with $^{15}N/^{14}N$ ratio of 3676.5×10^{-6}; and
- for S: the sulphur in a particular mineral (troilite, FeS) in the Canyon Diablo meteorite from Arizona, USA, written as CDT and having $^{34}S/^{32}S$ ratio of 45004.5×10^{-6}.

As an example, if in a particular sample we measure a $^{13}C/^{12}C$ ratio of 11276.5×10^{-6}, we are then able to apply Eq. 2.1 and calculate the $\delta^{13}C$ of that sample (relative to PDB), which will be 3.5‰.

1.2 Isotope Fractionation

As we discussed earlier, isotopes of the same element will have the same chemical characteristics; however, the difference in mass will cause them to have slightly different behaviour. This difference in behaviour—slight as it is—will cause different isotopes of the same element to partition, or segregate, between different compounds or phases: this phenomenon is called fractionation. We will examine exactly how and why this fractionation occurs, but first let us point out that the root cause for fractionation in equilibrium is the difference in bonding energy between the different isotopes of the same element, whereas the main cause for kinetic fractionation is the difference of diffusion rates between isotopes of the same elements. Both bonding energy and diffusion rate are a function of mass, which is the only difference between isotopes of the same chemical element.

Kinetic Fractionation

Kinetic fractionation is the dominant fractionation when rapid and nonreversible (unidirectional) reactions occur, in such cases as evaporation, diffusion, through many biogenic processes, or in any situation when equilibrium has not been fully reached. The main cause for kinetic fractionation is the difference in diffusion rates between isotopes of the same elements, which itself is due to differences in activation energy between two isotopes of different mass. Indeed, lighter compounds (those made of lighter isotopes) will

diffuse faster than heavier ones, as diffusion rate is function of mass. The result of kinetic fractionation is that the products of a reaction will always be lighter (depleted in the heavy isotope) relative to the reactants. As an example, a $^{12}C^{16}O_2$ molecule (mass of 44) will diffuse approximately 1.1% faster than the $^{13}C^{16}O_2$ molecule (mass of 45). This effect is more clearly noticeable during carbon fixation via photosynthesis in a plant, where organic matter is depleted in ^{13}C relative to air. Kinetic fractionations are typically very large and are often an order of magnitude larger than equilibrium fractionation. Importantly, the extent of kinetic fractionation is mostly independent of temperature.

Kinetic fractionation is often denoted by ε (epsilon) or Δ (delta), which represents the difference in isotopic composition between two phases: $\varepsilon_{A-B} = \delta_A - \delta_B$ (or $\Delta_{A-B} = \delta_A - \delta_B$). The extent of kinetic fractionation between different phases in a variety of reaction or phase changes is often established experimentally; however, the maximum magnitude of fractionation can be established theoretically.

Equilibrium fractionation

Equilibrium fractionation occurs when reversible reactions occur, i.e., when equilibrium between two phases has been achieved. This is often the case at high-temperature reactions, or when two phases interact for a long period of time. The extent of fractionation is lower, and the only factor at work is the difference in bonding energy between isotopes.

As an example, let us consider the oxygen molecule, O_2, which is a major constituent of the atmosphere and is an essential requirement for life on Earth. We will have three possibilities to combine the two main oxygen isotopes (^{16}O and ^{18}O) in the molecule: $^{16}O^{16}O$ (or $^{16}O_2$), $^{16}O^{18}O$, and $^{18}O^{18}O$ (or $^{18}O_2$). Of these, let us select the two most dissimilar, $^{16}O_2$ and $^{18}O_2$ and consider their characteristics. The first will have a mass of 32 and the second a mass of 36 (or a difference of 12.5%). This will result in the corresponding differences in density, and, in term, in bonding energy. The two O atoms in the molecule are subject to a repulsive and an attractive force, the combined effect of which holds the molecule together (Fig. 1.6). Because of the higher density and the greater attraction between the two ^{18}O, the "heavier" molecule $^{18}O_2$ will have a lower interatomic distance and result in higher dissociation energy than the $^{16}O_2$ molecule (e.g., 114.95 and 114.83 kcal/mole, respectively, at standard pressure and temperature). The dissociation energy is the energy that will have to be added to the system to dissociate the two atoms. In other words, the higher mass of ^{18}O results in higher dissociation energy, or in higher molecule bonding energy.

The same will be, of course, true for the different hydrogen molecules: $^{1}H_2$, $^{1}H^{2}H$, and $^{2}H_2$ will have dissociation energies of 103.2, 104, and 105.3 kcal/mole,

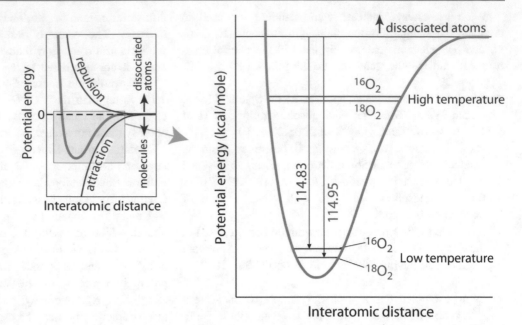

Fig. 1.6 The two different forces acting upon the two atoms of the O_2 molecule. The difference in potential energy between two oxygen molecules ($^{18}O_2$ and $^{16}O_2$) vary as a function of temperature

respectively, at standard pressure and temperature. These are significant differences: it would take approximately 2% more energy to dissociate the $^2H_2$ molecule than the $^1H_2$ one. These differences become immediately and practically observable when we consider O and H together in water. Let us consider water molecules made only with ^1H and those made only with ^2H: $^1H_2^{16}O$ and $^2H_2^{16}O$ (we will use the same oxygen isotope in the two cases). To start with, the mass difference between the two molecules is 11% (mass of 18 vs. mass of 20). The density will also vary, with 0.998 and 1.105 g/cm^3, respectively. The melting points will be 0 °C and 3.82 °C, respectively, and the boiling points will be 100 °C and 101.4 °C, respectively. These are very significant differences, and ones that will lead to noticeable fractionation between liquid water and water vapour, every single time evaporation or condensation occurs. In other words, the mass difference between H (and O) isotopes will cause significant isotopic fractionations between water and vapour every single time water changes state: this will become momentous when we consider isotopes and the hydrological cycle (Chap. 3).

Let us then recapitulate: because of their higher mass, heavier isotopes (of the same element) form stronger bonds than lighter isotopes, and thus produce more stable molecules. One major consequence of this effect can be found in the general fractionation rule stipulated by Jacob Bigeleisen (he of Manhattan Project fame) in 1965: "The heavy isotope goes preferentially to the chemical compound in which the element is bound most strongly." We can rephrase this to say that heavy isotopes tend to concentrate in phases with stronger chemical bonds. From now on, whenever in doubt,

let us return to this rule and apply it in any case, as it will always hold true.

Another consequence of the dissociation energy variations is that the extent of fractionation (how much fractionation will occur in a given chemical or physical process) will depend directly on the temperature at which this process occurs. If we have a second look at Fig. 2.7, we will realise that a second rule can be stipulated right away: the extent of fractionation will be higher at lower temperature, and vice versa. Again, this rule always holds true, and we would do well to apply it to any case of equilibrium fractionation that we deal with.

In order to quantify the extent of fractionation between two phases or compounds in equilibrium, we use the so-called fractionation factor, α (alpha). It is defined as the ratio between the isotope ratios in two compounds at equilibrium:

$$\alpha = \frac{\left(^{18}O/^{16}O\right)_{phase1}}{\left(^{18}O/^{16}O\right)_{phase2}}$$

From a practical point of view, and considering the definition of δ, some mathematical transformations (too tedious for us to go into here) allow us to approximate the difference between the isotopic composition of two compounds to $\ln\alpha$ * 1000:

$$\delta_A - \delta_B \sim \ln\alpha \times 1000$$

Crucially, and as explained above, this difference is always a function of temperature. As a result, if the isotopic compositions of two compounds at equilibrium are known, the temperature at which the reaction occurs can be found. On the other hand, if the isotopic composition of one of the two compounds is known, and the temperature of the equilibrium reaction is obtained by some independent method, it is possible to calculate the isotopic composition of the other compound.

For instance, let us consider quartz precipitating from water under equilibrium conditions. It has been established that the fractionation factor, $\ln\alpha \times 1000$, is 5‰ at 200 °C. Therefore, if the isotopic composition of quartz is −10‰, we can calculate that the water with which this quartz was in equilibrium with, had a δ of −15‰. If, alternatively, we have the isotopic composition of both quartz and water, we can in term calculate the temperature at which this equilibrium reaction occurs.

There are three main ways of determining fractionation factor s, α, as a function of temperature for any given couple of compounds:

- Theoretical, based on thermodynamic factors and constants, such as Gibb's free energy, enthalpy and entropy of the reaction, and the ideal gas constant. This method has no inherent temperature limitations and can be applied to any phase for which adequate thermodynamic data are available; however, it has limited accuracy because of the assumptions that are made during calculations.
- Semiempirical, based on the theoretical estimations described above, but also integrating laboratory experiments and natural samples.
- Empirical, based on laboratory experiments conducted in controlled temperature conditions. For instance, precipitation of a solid from a solution in equilibrium can be conducted at different temperatures, while taking samples of both the solution and the solid and measuring their isotopic composition.

1.3 Radiogenic Disintegration

By their very nature, isotopes that are not stable will tend to decay to stable ones. The process of becoming stable happens through radioactive disintegration: atoms will emit particles and energy until their nuclei reach the right combination of number of protons and neutrons necessary for maximum stability, on the stability line (Fig. 1.4). Radioactive isotopes are at a higher energy state and by emitting energy and particles; they will generate more stable isotopes found at a lower energy state, as demanded by the second law of thermodynamics.

The full list of naturally occurring radioactive isotopes with measurable abundances includes, with increasing mass,

^{3}H, ^{14}C, ^{40}K, ^{48}Ca, ^{50}V, ^{50}Cr, ^{70}Zn, ^{82}Se, ^{87}Rb, ^{96}Zr, ^{100}Mo, ^{113}Cd, ^{115}In, ^{123}Te, ^{124}Xe, ^{128}Te, ^{130}Te, ^{136}Xe, ^{138}La, ^{142}Ce, ^{147}Sm, ^{149}Sm, ^{150}Nd, ^{152}Gd, ^{174}Hf, ^{176}Lu, ^{180}Ta, ^{183}W, ^{184}W, ^{184}Os, ^{186}Os, ^{187}Re, ^{190}Pt, ^{232}Th, ^{235}U, and ^{238}U. (There are many other, very short-lived or synthetic, radioactive isotopes that are of interest mostly to nuclear physicists and are beyond the scope of this book.) Some elements (e.g., H, C, K) have only one minor radioactive isotopes (with others being stable), whereas some elements (e.g., Th, U) have only radioactive isotopes. Notice that the great majority of radioactive isotopes have masses above 56 (^{56}Fe, the most stable of nuclides), which reflects their formation in conditions of disequilibrium: the further we go from ^{56}Fe, the fewer stable isotopes we will encounter.

Simplistically, there are three main groups of radioactive decay modes (Fig. 1.7): (1) decay with emission of nucleons; (2) beta (β) decay; and (3) transition between states of the same nucleus, also known as gamma decay (γ). Let us consider them briefly here:

(1) Decay with emission of nucleons. These include emission of a proton from the nucleus, emission of a neutron, double proton emission, double proton emission, emission of a series, or clusters of protons and neutrons, or the most common, emission of a particle of two protons and two neutrons, or α particle (α radiation). Alpha radiation has the lowest energy and can be completely stopped by skin or a sheet of paper, thus representing little danger to human health.

(2) Beta (β) decay. This mode of disintegration involves the transition of one neutron to one proton (β⁻, beta minus), with the emission of an electron and energy. This also happens through the transition of one proton to one neutron (β⁺; emission of energy one and

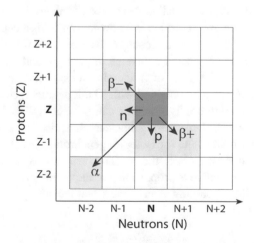

Fig. 1.7 Simplified representation of the most common disintegration modes: emission of a neutron, emission of a proton, conversion of a proton to a neutron, or a neutron to a proton, or emission of an α particle, consisting of two protons and two neutrons

positron; or with electron capture (EC). In this case, one electron from the innermost electron shell is captured and, combined with one proton from the nucleus, produces one neutron and liberates energy. We can also have double β^- and double β^+ decays. Beta decays also have low energy, albeit higher than α, and can be stopped by clothing or an aluminium sheet, thus representing a limited danger to our health.

(3) Gamma (γ) radiation, on the other hand, is a very high-energy radiation and is emitted when an isotope transitions from an excited, high energy, to a lower energy (and more stable) state, coupled with the emission of the highest energy photons. Because of this very high energy, gamma radiation represents an acute danger to human health and can be stopped only by very substantial shielding, such as a several metres thick concrete wall, a several centimetres-thick lead lining, or a combination of both.

Some isotopes have a simple, one-step disintegration path, such as ^{147}Sm, which disintegrates to the stable ^{143}Nd by emitting one α particle and energy (hence the mass difference of 4). Others, such as U (^{238}U and ^{235}U), have a protracted and complicated disintegration schemes: for instance, ^{238}U disintegrates to the stable ^{206}Pb following a series of 8 α and 6 β^- emissions, involving 13 short-lived intermediate isotopes (each of which is thus both radiogenic and radioactive). Yet others have more than one disintegration scheme: as an example, the minor potassium isotope, ^{40}K, will follow four distinct paths to stability: (1) 11% of ^{40}K will undergo electron capture to the excited state of ^{40}Ar, which will then emit γ radiation and go to the fundamental state of ^{40}Ar; (2) 0.16% will undergo direct electron capture to ^{40}Ar; (3) 0.001% will undergo β^+ decay to ^{40}Ar; and (4) 88.8% will undergo β^- decay to ^{40}Ca.

Importantly, radioactive decay (or disintegration) occurs at a constant speed, which is known, with a high degree of precision, for each radioactive isotope. This has very important and useful significance in geology and archaeology, where the age of minerals, rocks, and artefacts, respectively, can be obtained with a high degree of certainly. Let us examine briefly how this can be accomplished.

Rutherford, a pioneering nuclear physicist and also the godfather of absolute (or isotopic) geochronology, stated the fundamental principle thus: "The rate of decay of a radioactive nuclide is proportional to the number of atoms of that nuclide remaining at any time." Mathematically, this can be expressed with

$$-\frac{dN}{dt} = \lambda N$$

where t is the time elapsed, λ is the disintegration constant (specific to each radioactive isotope, as shown in Appendix A), and N is the amount of radioactive parent remaining. The amount of daughter isotope coming directly from the disintegration of the parent (D*; the asterisk denotes its provenance from radioactive disintegration) is the difference in amounts between the original parent at time 0 (N_0) and the remaining parent (N):

$$D^* = N_0 - N$$

Combining those two equations and integrating them will give us

$$N_0 = Ne^{\lambda t}$$

When we replace N_0 in this equation, we obtain the relationship between the remaining amount of parent and the amount of daughter isotopes, as function of time:

$$D^* = N\left(e^{\lambda t} - 1\right)$$

This equation can be then solved for time:

$$t = \frac{1}{\lambda}\ln(D^*/N + 1)$$

This is the general equation used in geochronology for any isotopic parent–daughter system: if we know the amounts of the daughter and parent isotope (expressed as number of atoms) at present, the value of the disintegration constant, and if we follow some fundamental assumptions (into which we will go in greater detail later), we can calculate the age of any mineral, rock, or historical artefact, with a high level of certainty.

Further Reading

Many books describe the basic principles of isotopes and isotope fractionation, some in greater detail, others in terms easier to understand. Here is a short selection of these:

Principles of Stable Isotope Distribution, R.E. Criss, Oxford University Press, 1999, ISBN 978-0-19-511775-2. Of greatest use for our purposes here are Chapters 1, 2, and 4, as well as Appendix A3.

Stable Isotope Geochemistry, J. Hoefs, Springer-Verlag, 1997, ISBN 3-540-61126-6. Chapters 1 and 2 are the most significant to us.

The Elements, P.A. Cox, Oxford University Press, 1990, ISBN 0-19-855298-X. A very interesting little book, the most useful chapters being 1, 3, and 6.

Stable Isotope Geochemistry: a Tribute to Samuel Epstein, H.P. Taylor, J.R. O'Neil, and I.R. Kaplan, Editors. The Geochemical Society Special Publication #3, ISBN 0-941809-02-1. Of particular interest to us is Part A, which discusses isotopic fractionations.

Isotopes: Principles and Applications, G. Faure and T.M. Mensing, Wiley, 2005, ISBN 978-047-138437-3.

Questions

- What are isotopes defined by? What is the single most important characteristic of an isotope? What is the difference between the different isotopes of the same chemical element?
- When, where, and how are isotopes formed?
- What is the line of stability? Why is it positioned in its particular location? What happens to isotopes that are situated away from it?
- How do we calculate isotopic composition? Relative to what do we calculate it?
- What is fractionation? What types of fractionation are there and what factors affect them? How do we calculate fractionation?
- What is radiogenic disintegration? What modes of disintegration and what types of radiation are there?

It is important, before we dive into the many applications of isotopes in the natural environment sciences, to consider how exactly we analyse our samples: what analytical techniques we use, how we prepare our samples, and how we interpret our results. It will be easier to understand the specific applications of isotopes with this knowledge under our belts. We will also devote some space to the questions of how we prepare our samples and how we treat and interpret the analytical data.

2.1 Mass Spectrometry

As we discussed earlier, the main defining characteristic of any isotope is its mass, which is given by the sum of its number of protons and number of neutrons in its nucleus. If we want to analyse the isotopes of the same chemical element, the only way to do it would be to have a tool that can distinguish them by their mass, and this tool is the mass spectrometer. In its fundamental principle, this is a simple and straightforward machine, and it works thus (and in high vacuum; Figs. 2.1 and 2.2).

Firstly, we need to separate different elements from each other by breaking the bonds that hold them together; namely, separating atoms from the molecules that they were part of. This part is often done before the sample reaches the mass spectrometer, or can be the first thing that happens there, in conjunction with the second step.

Secondly, we ionize the atoms. To do this, we subject the atoms to a cloud of electrons generated by a small coiled filament—not too dissimilar to that found in incandescent light bulbs, and often made of tungsten, too—to which we have applied an electric current. To place the cloud of electrons and the sample atoms in contact, we have two small plates, charged contrastingly, that move the cloud of electrons in the middle of the source, where we have injected the sample.

Thirdly, we accelerate the sample ions. This is a critical step, as we want to make sure that the speed at which they are flying is constant and well defined since any variation, even weak, will make measurement impossible. The acceleration is accomplished by a couple of plates that are charged contrastingly, with a potential of a few thousand volts between them. The ions will be repulsed by the repulsion plate and attracted by the acceleration plate. The latter has a small slit in it through which the ions will fly, straight and at high speed, down the flight tube. The ion beam thus generated, made up of all the ions present in the sample, will be shaped and focused by a few couples of plates that are charged contrastingly and act as electrostatic lenses.

Now comes the critical part: the ion beam reaches a strong magnetic field that is generated by a large electromagnet. This makes the ion beam deviate from its straight trajectory and curve; the flight tube also curves to accommodate the deviation. Importantly, the heavier isotopes will be that little bit more difficult to deviate from the straight path than the lighter isotopes, due to their higher momentum. After the ion beam passes through the magnetic sector there will be not one ion beam, but several, each made of ions of the same mass, at all other conditions (in particular the acceleration voltage) maintained stable. The heavier isotopes will follow a slightly less curved path and the lighter isotopes will be on a more curved path. In other words, the kinetic energy of different isotopes, directly function of their mass, will allow for the magnetic sector to separate the isotopes of the same element (Fig. 2.3).

The ion beam is now made up of only isotopes of the same mass, as the heavier and lighter ones are deviated too little or too much, respectively, and hit the walls of the flight tube. The final step of the process is when this chosen beam, containing only the mass we are interested in, reaches the detector, where the number of ions is counted. There are different types of detectors, but generally the main idea for

© Springer Nature Switzerland AG 2020
P. Alexandre, *Isotopes and the Natural Environment*, Springer Textbooks in Earth Sciences,
Geography and Environment, https://doi.org/10.1007/978-3-030-33652-3_2

Fig. 2.1 The simplified schematic representation of a mass spectrometer, with the main parts: the source, the magnetic sector, and the detector (SEM and Faraday Cage are shown). The magnetic field is adjusted in such a way to allow only an ion beam made of ions of specific mass to reach the detector

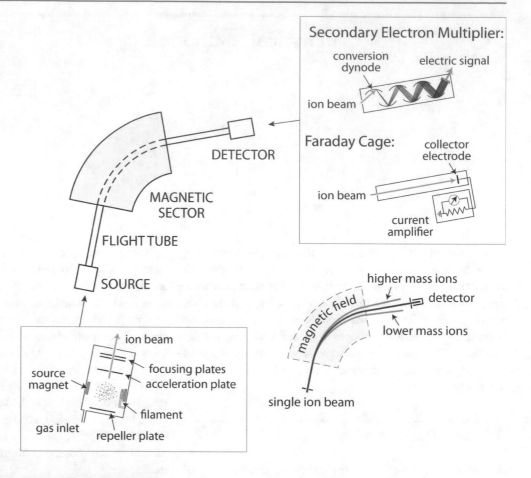

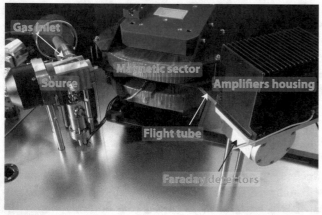

Fig. 2.2 An example of a modern mass spectrometer, the ARGUS multi-collector mass spectrometer specifically designed for $^{40}Ar/^{39}Ar$ geochronology, with the main elements visible. It has five Faraday detectors; the source and the detector housing are made out of solid hollowed out blocks of steel. The amplifier housing operates under vacuum, as well

At any point, we can select which mass we want to measure by modifying the operating conditions of the mass spectrometer. There are two ways to do this: (1) vary the acceleration voltage and give the primary ion beam different speeds, while keeping the magnetic field stable (in this case we will use a permanent magnet), and (2) keep the accelerating voltage and the primary ion beam speed constant, but vary the intensity of the magnetic field. This can be accomplished by varying the electric current that we apply to the electromagnet we use. Both methods are widely used; the former is predominantly used for light stable isotopes, whereas the latter is more common and can be used for any isotope of any mass.

So far, things are simple and clear, but this will not last for very long, as many complications and variations exist. These are necessary mostly because of the great variability of the natural samples that we analyse. Let us now consider some of these complications.

Type of Source

The source described above—a very common one—is called gas source, as we introduce the sample in the form of gas (Fig. 2.4). This is a stable source that is easy to use and maintain, so we often convert the sample to gas by following

all of them remains the same: the amount of ions arriving in the detector is precisely and directly proportional to an electric current generated by it, which can then be measured using a simple voltmeter.

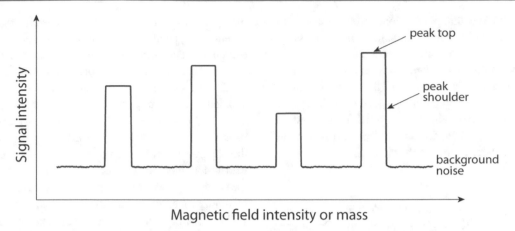

Fig. 2.3 Idealised peak scan, showing the different elements of a well-formed peak. As the intensity of the magnetic peak is progressively increased, ion beams corresponding to different masses will reach the detector and be counted (typically in counts per second). The specific magnetic field intensity for each mass is known and allows us to select the mass to be analysed; if more than one mass needs to be analysed, the magnetic field intensity is modified accordingly: this method is called peak jumping

Fig. 2.4 Two different views of a modern gas source, very compact and mounted on a single flange; it is used in the mass spectrometer shown in Fig. 2.2. The source magnets, the exit slit, the focusing and accelerating plates (as per the schematics on Fig. 2.1), and the electric connectors for the different components are visible

the sample preparation (discussed below). Another source is designed to use solid samples: a small amount of sample is deposited (often precipitated from solution) on a small tungsten filament which is then placed in the source. When the sample is heated through the application of an electric current to the filament, the sample bonds are broken and the atoms are ionized. Unsurprisingly the process is called thermal ionization, and the analytical method is called thermal ionization mass spectrometry (TIMS).

TIMS has been used for decades and is still in use, but tends to be difficult and slow, as placing the samples on the filament and then placing the filament in the source is not the simplest of procedures. Another method has been developed to use with solid samples or with solutions, and this is to heat the sample with plasma. The sample will be in the form of a nebulized solution (a fine mist of the solution into which our sample has been placed) or of fine grains (well under 1 μm) produced by a laser beam hitting the surface of the sample.

Nebulization happens immediately before the plasma, but the laser ablation happens further upstream, in a laser installation coupled to the mass spectrometer. The fine grains of sample are carried to the plasma suspended in gas (typically noble gas, so as not to interfere too much with our analysis). The plasma itself is produced by a varying magnetic field that changes its polarity at radio frequency; the magnetic field is generated by an electric current that goes through a copper coil and changes its polarity at the same frequency. As all the particles are forced, by the magnetic field, to turn this way and then the opposite way at radio frequency, they heat up by friction, reaching temperatures of up to 10,000 °C. Because of the way it is generated, the plasma is called inductively coupled plasma (ICP), and the method, ICP-MS.

Very few bonds will survive the plasma conditions, so we end up with individual atoms which are also ionized: the plasma "torch" serves as our source. One major difference from the gas source is that plasma ionization occurs at atmospheric pressure. To introduce our sample in the mass spectrometer, working in high vacuum, we use an interface which is basically two small plates with a tiny hole in each: the ions will be sucked in the high vacuum of the mass spectrometer where they will be accelerated and analysed by the magnetic sector. The method, laser ablation ICP-MS (LA-ICP-MS), or solution ICP-MS (S-ICP-MS) is very powerful and fairly common, particularly in the Earth sciences.

Resolving Power

There are many isotopes of different elements that have the same mass, such as ^{40}K and ^{40}Ca, or ^{50}V and ^{50}Cr, and many others. How can we separate these from each other in the magnetic sector? To start with, their actual masses are slightly different: ^{40}K has a mass of 39.9640, whereas ^{40}Ca has a mass of 39.9626; this is a difference of 0.0014. The individual peaks corresponding to the two isotopes would have to be very narrow to be able to see two peaks here and not one. How narrow? A simple calculation allows us to estimate it: we divide the mass (M, 40 in this case) by the difference in mass (ΔM, or 0.0014) and come to a required resolving power of approximately 28,500. However, even the most complex, large, and powerful mass spectrometers are not able to achieve such resolving power: in practical terms, mass spectrometers have mass resolution (a synonym for resolving power) of under 10,000. And what about ^{50}V and ^{50}Cr? Using their actual masses (50.9440 and 49.9461, respectively) we come with the required mass resolution of approximately 24,250, yet this cannot happen. So, what are we able to do? Simply, we separate the elements before introducing them in the mass spectrometer, by using their different chemical characteristics. That, in itself, is not always easy, but is the only way we are able to go about it.

Type and Number of Detectors

There are two major types of detectors: the Faraday collector (also known as Faraday cage) and the secondary electron multiplier and its derivatives, the channeltron and the microchannel plate.

The Faraday cage is, in itself, a very simple device: it consists of a small metallic box with an opening on one side and a wire attached to it on the other side—and of very little else. As an ion hits the inside of the box, it will mobilize one or more electrons, thus creating a very weak electric current. As this current is too weak to be detected, it must be amplified by an amplifier (consisting of a single resistor), which typically increases the current 10^{11} or 10^{12} times, thus making it easily measured by an amp-metre. The number of ions penetrating the Faraday cage is proportionate to the electrical current detected.

The secondary electron multiplier (SEM), is a similarly simple device, consisting of several small curved copper plates arranged in two rows facing each other. An electric potential—typically a few thousand volts—is applied on each plate. The very first plate, called conversion dynode, is the one that the incoming ion beam will hit, generating a small electron beam. That beam will hit the second plate, situated on the opposite row that generates even more electrons, due to the voltage applied to the plates. This process will go on: more electrons being generated each time the electron beam hits a subsequent plate, thus producing a measurable electric potential at the last plate, which is then measured by the volt-metre. In this case, the number of incoming ions will be proportional to the voltage measured.

The channeltron is a development of the secondary electron multiplier and consists of a curved and constantly narrowing tube (think a portion of a seashell). Electron multiplication occurs continuously as electrons bounce inside it. Finally, the microchannel plate works on the same principle as the SEM, but instead of opposing curved plates, we have a series of hollow glass capillaries within which electron multiplication takes place.

The two most common detectors are the Faraday cage and the SEM. They both have a range of advantages and disadvantages such as price, durability, ease of use and maintenance, size, etc. Importantly, the SEM is typically one order of magnitude more sensitive than the Faraday cage and is therefore preferred when very low concentrations are expected. On the other hand, the Faraday cage is cheaper, smaller, and easy to maintain, which are all important characteristics. As a result, both detector types are commonly used in mass spectrometry and which one is preferred will depend on the specific applications and the analytical needs.

So far, we have considered mass spectrometers with only one detector, but this is not always the case: indeed, there are possibly more mass spectrometers with more than one detector, which we will call multi-collector. Such an arrangement is very important, as it allows us to measure more than one isotope at a time.

2.2 Accessories and Add-ons

In most cases, the sample will be prepared off-line, which is to say, on a separate analytical installation, and then transported to the mass spectrometer in a glass vial in gaseous form. The vial is coupled to the mass spectrometer and the sample gas is introduced and analysed (in conjunction with a standard reference gas), relative to which the analytical results are calculated. There is a great variety of off-line sample preparation methods, which will be briefly discussed below, but also a series of online add-ons, which prepare the sample and introduce the gas to be analysed directly in the mass spectrometer. The good news is that these add-ons

Fig. 2.5 Simplified schematic diagram of an element analyzer. The most important component is the gas chromatography column, which separates molecules by size and allows for their individual isotope analysis

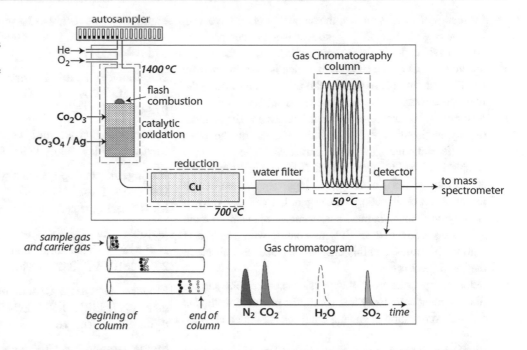

work automatically, and thus make our lives much easier. Of these the most important and very commonly used is the Thermal Combustion Elemental Analyzer (TC-EA; Fig. 2.5).

The elemental analyzer consists of (1) an autosampler, which is basically a flat aluminium disc with holes in it. Each sample is placed in a hole, and as each hole comes above an opening, the sample falls into the (2) combustion chamber. Here, the sample is combusted in excess of oxygen at a very high temperature (typically above 1400 °C), and all elements are oxidized. Then, the resulting gas is reduced in the presence of copper in (3) the reduction chamber and then penetrates in (4) the Gas Chromatography (GC) column. This is a coil of a very fine tube (internal diameter typically between 0.1 and 0.5 mm), typically between 10 and 60 mm in length, and coated internally with particular material (polyamide, fused silica, or a liquid phase that has been bonded to the inner surface). As the gas sample flows through this tube, the different molecules will travel at different speeds, with smallest ones arriving at the other end before the larger ones. This allows for an effective (and inexpensive) way to separate by size the molecules present and analyse their relative amounts by (5) a small electronic detector, before introducing the gas in the mass spectrometer and analysing the isotopic composition of each molecule. This is a very common and powerful method and is widely used for analysing the isotopic composition of a range of natural materials.

2.3 Sample Preparation

A significant range of sample preparation methods is available; however, any specific method used will depend on the material analysed and the isotopes measured. These methods also range in difficulty, cost, and duration, from cheap, fast and easy to slow, costly, and difficult, and any combination in between. Let us have a quick and necessarily superficial look at the main methods, organized by chemical element, for the five most commonly isotopically analysed elements in natural sciences studies.

- Hydrogen. Independent of the kind of material analysed (water, solids, or organic matter), hydrogen is extracted by high temperature reduction in the presence of uranium or copper. The hydrogen thus extracted is then analysed in the mass spectrometer, either directly or off-line.
- Carbon. For carbon in carbonate minerals, the sample is dissolved off-line in phosphoric acid and the liberated CO_2 is then transferred to the mass spectrometer to be analysed. For organic matter, the sample is prepared by an online elemental analyzer.
- Nitrogen. The sample can be gas or organic matter. If gas, the sample is directly introduced in the mass spectrometer and analysed there. If the sample is organic matter, the sample is prepared by an online elemental analyzer.

- Oxygen. For water samples, the sample is placed in a vial, in contact with pure CO_2 of known isotopic composition. After a while (typically a day), during which the vial is agitated, the CO_2, the oxygen in which is now isotopically equilibrated with the oxygen in the water, is analysed in the mass spectrometer. Organic matter is analysed after first preparing it by an online elemental analyzer. The analysis of oxygen isotopes in rocks and minerals, on the other hand, is difficult, slow, and expensive. The powdered sample is placed in contact with BrF_5 (a highly corrosive and explosive acid) overnight at high temperature, which has the effect of liberating the oxygen. In rare cases, oxygen is analysed online in a mass spectrometer, but is much more commonly converted to CO_2 in contact with graphite at high temperatures, and this CO_2 is then analysed off-line in the mass spectrometer.
- Sulphur. Irrespective of the physical state of the sample (though most commonly it is solid), the preferred method is thermal combustion elemental analyzer, described above.

There are a great many other analytical methods and elements whose isotopes are routinely analysed, but there is little room here to go into any detail about them.

Further Reading

Unfortunately, the analytical methodology of isotopes, from sample collection and preparation, to analysis, and to data treatment and interpretation, is rarely described in a single book, which leaves us to look into a great many papers in order to collect the information required. Further, this topic is best learned in the laboratory, by actually conducting analytical work. However, there are a few sources that we can peruse and learn from:

Handbook of Stable Isotope Analytical Techniques, Volume I, P.A. de Groot, Elsevier, 2004, ISBN 978-0-444-51114-0.

Handbook of Stable Isotope Analytical Techniques Volume II, P.A. de Groot, Elsevier, 2004, ISBN 978-0-444-51115-7.

Isotopes: Principles and Applications, G. Faure and T.M. Mensing, Wiley, 2005, ISBN 978-047-138437-3

Radiogenic Isotope Geology, A.P. Dickin, Cambridge University Press, 2005, ISBN 0-521-82316-1. Chapter 2, Mass Spectrometry, is the chapter of interest to us.

Isotopic Analysis: Fundamentals and Applications Using ICP-MS, F. Vanhaecke and P. Degryse, Editors, Willey, 2012, ISBN 978-3527328963.

Modern Analytical Geochemistry, R. Gill, Editor, Longman, 1997, ISBN 0-582-09944-7. We will focus on Chaps. 8–10, 12, 15, and 16.

Stable Isotope Geochemistry, J. Hoefs, Springer-Verlag, 1997, ISBN 3-540-61126-6. Chapter 1 contains some important pointers.

Using Geochemical Data, H. Rollinson, Pearson Education, 1993, ISBN 0-582-06701-4. The relevant chapters are Chaps. 6 and 7.

Questions

- Which characteristic of isotopes do we use to measure their concentrations and ratios? What is the corresponding analytical instrument called and how does it work? What are its main components and how do they work?
- What is resolving power?
- What sort of accessories and add-ons are there?
- What are the main types of sample preparation for the major stable isotopes?

Part II
Applications of Isotopes in the Natural Environment

Stable Isotopes and the Hydrosphere

3.1 The Importance of Water

There is little need to elaborate the importance and significance of water for all forms of life on Earth (including us) since life on Earth is water based. In addition, a very significant portion (approximately 71%) of the planet's surface is covered with water in the form of oceans and lakes, which are also host of a vast amount of aquatic plants and animals. Water also participates in and interacts with the lithological sphere (or the geosphere), the biosphere, and the atmosphere. All in all, a very complex and extensive web of interactions affect water in its global cycle, and each of these will modify its isotopic composition to some extent. By studying the isotopic composition of the two main components of water, hydrogen, and oxygen, we can deduce the processes that took place and the conditions at which they occurred, and this is what we will consider in the first section of this chapter.

3.2 Water on Earth

As mentioned, the largest proportion of terrestrial water by far —at nearly 97%—is found in oceans, as seen in Table 3.1. This is followed by ice and groundwater, which combine for a total of around 3.5%, and all other reservoirs, which represent much less than one percent.

Water has two main components—hydrogen and oxygen —with two stable isotopes for the former (^{1}H and ^{2}H) and three for the latter (^{16}O, ^{17}O, and ^{18}O; for the purposes of this book, only ^{16}O and ^{18}O will be considered, nor will we consider the radioactive ^{3}H). ^{1}H is by far the dominant hydrogen isotope, with an abundance of 99.984% (and 0.016% for ^{2}H), and ^{16}O is the major oxygen isotope, with an abundance of 99.76% (and 0.20% for ^{18}O). When the exact masses of these isotopes are considered, the mass difference between ^{1}H and ^{2}H is 99.8%, and is 12.5% for ^{16}O and ^{18}O. We will notice that the mass difference for hydrogen is 8 times higher than the mass difference for oxygen. This is not an idle observation and its significance is apparent when we realise that, during any physiochemical process, ^{2}H and ^{18}O should ideally separate from the same phase by a factor of 8 (Table 3.1).

3.3 Fractionations During Evaporation and Condensation

As with any other stable isotope, we will follow the same δ (delta) notation, expressing how much the heavy isotope (in this case ^{2}H or ^{18}O) is enriched or depleted in a sample relative to a standard, as described in Chap. 2. The same applies for the fractionation factor, α, defined as the ratio of isotope rations in two compounds (e.g., liquid water and water vapour):

$$\alpha_{L-V} = \left(^{18}O/^{16}O\right)_L / \left(^{18}O/^{16}O/\right)_V$$

The fractionation factor α can be related to the delta notation:

$$\delta_L - \delta_V \sim ln\alpha_{L-V} \times 1000$$

On the other hand, the fractionation factor is a function of temperature: the higher the temperature, the lower the extent of fractionation. This inverse relationship can follow different forms, and is linear in the case of evaporation and condensation:

$$ln\alpha_{L-V} \times 1000 = a/T^2 + b$$

where T is the temperature, in degree Kelvin. If we consider the fractionation for hydrogen and for oxygen isotopes during evaporation at different temperatures (Table 3.2), we will observe that it is 8 times larger for ^{1}H than for ^{18}O at temperatures between 20 and 25 °C, as predicted theoretically: the difference in mass is 8 times larger (2 and 1 AMU) for hydrogen than it is for oxygen (18 and 16 AMU).

Let us now look at a simplified version of the global water cycle, in which we consider only evaporation (mostly

P. Alexandre, *Isotopes and the Natural Environment*, Springer Textbooks in Earth Sciences, Geography and Environment, https://doi.org/10.1007/978-3-030-33652-3_3

Table 3.1 The major water reservoirs on Earth and their hydrogen and oxygen isotopic composition. Notice the highly variable composition of each water type, with the exception of the ocean, reflecting the diversity of processes affecting them. The isotopic composition of ocean water is 0‰ and 0‰ by definition, as we use it as standard

Reservoir	Volume (10^9 km^3)	Volume (%)	δ^2H (‰)	δ^{18}O (‰)
Oceans and seas	1.34	96.6	0 ± 5	0 ± 1
Ice caps, glaciers, permanent snow	0.0256	1.843	-230 ± 120	-30 ± 15
Groundwater	0.0234	1.684	-55 ± 60	-4 ± 8
Lakes and swamps	0.000188	0.0135	-45 ± 60	-6 ± 6
Atmosphere	0.0000123	0.00089	-150 ± 80	-20 ± 10
Rivers	0.0000021	0.00015	-45 ± 60	-6 ± 6
Biological water	0.0000011	0.00008		
Total	*1.3892*	*100*		

Table 3.2 Fractionation factors for H and O during evaporation at different temperatures. At lower temperatures the α_{1H}/α_{18O} ratio will increase to about 8.7, whereas it will decrease to about 7.5 at higher temperatures

Temperature, °C	α^1H$_{V/L}$	α^{18}O$_{V/L}$
0	−101.0	−11.55
5	−94.8	−11.07
10	−89.0	−10.60
15	−83.5	−10.15
20	−78.4	−9.71
25	−73.5	−9.29
30	−68.9	−8.89
35	−64.6	−8.49
40	−60.6	−8.11

from the oceans) and precipitation. We will also further simplify the situation by assuming (which is not that removed from reality) that most evaporation occurs in the intertropical region (between latitudes of approximately 23.5° S and 23.5° N), and that the most precipitation occurs in the temperate, cooler regions closer to the poles. This situation implies a generalized and simplified global movement of atmospheric water from equatorial (mostly evaporation) to polar (mostly precipitation) regions, corresponding to the predominant winds.

As water evaporates, the isotopic composition of the water vapour (in the atmosphere) will be depleted in the heavy isotope, which will remain preferentially in the water phase (in the ocean). Thus, the initial vapour phase will have a δ^{18}O of approximately −14‰, a value obtained empirically. As this moisture travels away from the equatorial regions and a first precipitation occurs (Fig. 4.1), condensation will result in liquid water that is enriched in ^{18}O (at a δ^{18}O of −4‰) and a remaining vapour phase further depleted in ^{18}O (at −16‰), reflecting a fractionation factor during precipitation ($H_2O_{vapor} \rightarrow H_2O_{water}$) of 10‰ (Fig. 3.1). When a second precipitation occurs further away from the equator, in cooler regions, the liquid phase will be depleted in ^{18}O relative to the first precipitation, as if formed from an isotopically lighter vapour.

Each subsequent precipitation is isotopically lighter than the previous one (Fig. 3.1), as the process of condensation repeatedly occurs. This will be true for both hydrogen and oxygen isotopes. If we plot all observed precipitation waters (called *meteoric* waters) on a δ^{18}O versus δ^2H diagram (Fig. 4.2), all observations will fall on approximately the same line, called Global Meteoric Water Line. This line is described by the theoretical equation

$$\delta^2H = 8\delta^{18}O + 10$$

The slope of the line, 8, corresponds to the fact, noted earlier, that hydrogen fractionation will be 8 times more extensive than the oxygen fractionation. This is due to the mass differences between the two isotopes of each element (100% for H, 12.5% for O, and a ratio of 8 between the two). The extent of the observed fractionation can be predicted based on the mass differences between isotopes, and thus we call this *mass-dependent fractionation*, which is by far the most common type. Mass-independent fractionation can also occur, but is very rare and will be discussed later, in Chap. 6.

The shift between δ^2H and δ^{18}O of 10‰ is called deuterium excess and corresponds to the difference in kinetic fractionation between the two elements: let us not forget that evaporation is not an equilibrium process and fractionation during evaporation will be partially kinetic.

Fig. 3.1 Simplified evaporation–precipitation water cycle and the evolution of oxygen isotopic composition. The initial water vapour will have lighter oxygen than the ocean water. Each precipitation will preferentially remove heavier oxygen resulting in progressively lighter oxygen as we move away from the source of moisture

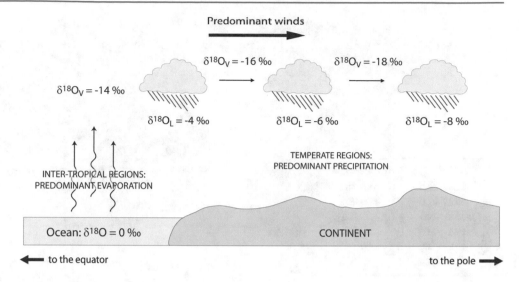

If all actual observations of meteoric water isotopic compositions are considered together, we will notice a certain variation in both the slope of the line and the deuterium excess, as seen in Fig. 3.2 and reflected in the following equation

$$\delta^2H = (8.17 \pm 0.06)\delta^{18}O + (10.35 \pm 0.65)$$

In reality, significant variations are observed for both the slope of the line and for the deuterium excess, due to secondary factors that we will review shortly. The slope of the line varies from 5.5 to 8.1 for land-based observation stations and from 2.8 to 7.1 for marine stations. Deuterium excess varies from 1.4 to 11.5 for land-based observation stations and from −1.1 to 6.6 for marine stations.

The most significant of the secondary factors is temperature. We can observe—for Norther America and Europe—a clear relationship between the mean annual air temperature and isotopic composition of meteoric waters:

$$\delta^{18}O = 0.695T - 13.6\%$$

and

$$\delta^2H = 5.6T - 100\%$$

These relationships stem directly from the fractionation factors, the extent of which is, as we have considered earlier, a function of temperature.

Several other factors—often interrelated—affect the isotopic composition of meteoric waters, and are mostly defined empirically:

- Altitude effect (approximately $2 \pm 1\%$/km for $\delta^{18}O$). A very good case of altitude effect has been observed in the Canadian Rocky Mountains, with $\delta^{18}O$ of rain at sea level approximating −11‰ and $\delta^{18}O$ of precipitations at 3.5 km above sea level approximating −19‰.
- Latitude effect (approximately $0.002 \pm 0.001\%$/km for $\delta^{18}O$);
- Distance from the coast effect (rain-out effect). As the predominant air mass carrying moisture moves from west to east across Europe, δ^2H varies from approximately −45‰ in England to about −55‰ in France, and to about −65‰ in Romania.
- Amount effect. The larger the amount of precipitation, the lower the meteoric waters $\delta^{18}O$ and δ^2H are. The meteoric water $\delta^{18}O$ recorded by tropical island stations varies from around 0‰ for small precipitations (under 50 mm) to approximately −5‰ for large precipitations

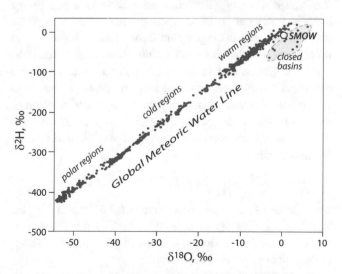

Fig. 3.2 Hydrogen and oxygen isotopic compositions of precipitation (meteoric) waters from different parts of the world. Closed, strongly evaporative basins, and the Standard Mean Ocean Water (SMOW) are also shown. As more and more vapour water undergoes condensation and precipitates as rain (or snow, closer to the poles), each subsequent precipitation will have lower δ^2H and $\delta^{18}O$, always falling on or close to the Global Meteoric Water Line (data from Dansgaard [2], Gat [3], Yurtsever [5])

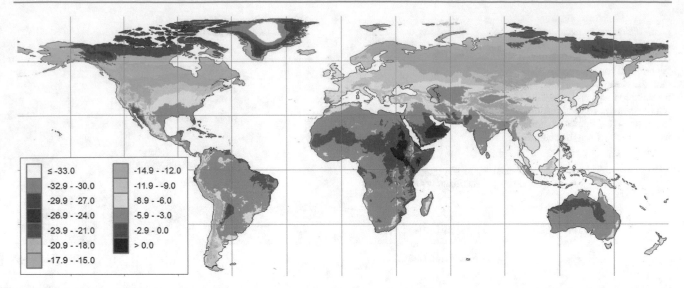

Fig. 3.3 Interpolated global $\delta^{18}O$ composition of meteoric waters. All the effects affecting the isotopic composition of precipitation are clearly visible, such as the main equator–pole variation, the altitude effect (e.g., the Andes of South America, Himalayas), and the distance from the coast (e.g., Europe, South America). Reprinted with permission from Terzer et al. [4]

(larger than 300 mm). This is due to more complex phenomena in larger precipitations, such as more intense vertical mixing (due to more significant vertical temperature variations), a larger proportion of the evaporation of falling drops, and the size of the drops (smaller drops exchange isotopically with near-surface moisture more than larger drops).

The major consequence of the effects described here, from the main variation observed in Fig. 3.2 to all secondary effects, is that each location on Earth will have a specific and known meteoric water isotopic composition (Fig. 3.3). If you sample a river in the Yukon, or near Cape Town, or in Patagonia, the isotopic compositions will be very distinct. In other words, each surface water that we analyse will tell us quite clearly where precipitation took place.

3.4 Rayleigh Distillation

There are three possible ways in which any given system can evolve with regards to evaporation and precipitation. These include open system, where the evaporated water is constantly removed (and not in isotopic equilibrium with liquid water), closed system, where the evaporated water remains in contact and thus in isotopic equilibrium with the liquid water; and steady-state system, where the outgoing flux is replaced by an equal amount of inflow.

The evaporation part of the simplified global evaporation–precipitation cycle that we examined above corresponds to a steady-state situation: ocean water is constantly

evaporated, but also constantly replenished by rivers and its volume remains fairly constant. Ocean water's isotopic composition remains constant: on one hand it becomes heavier due to evaporation, but on the other hand, it is replenished by isotopically lighter river waters.

Contrastingly, the precipitation part of the global water cycle corresponds to an open system: liquid extracted from the vapour in a particular air mass leaves the system. This process can be formalized and quantified using the co-called *Rayleigh distillation*. Rayleigh distillation is a process of isotopic fractionation in an open system, where one phase (the vapour water in evaporation from a finite reservoir, or in inverse, the condensate from a finite reservoir) is removed from the system and is no longer in contact and isotopic equilibrium with the other phase. Importantly, Rayleigh distillation allows us to quantify the isotopic composition of the two phases at any point of the evaporation process, as a function of the proportion of water evaporated, and following this equation:

$$R_t = R_0 f^{(1-\alpha)}$$

where R_t and R_0 are the isotopic ratios (or $\delta^{18}O$ and δ^2H, for liquid or for vapour) at times t or 0, f is the fraction of vapour or liquid remaining, and α is the fractionation factor—itself dependent on temperature.

In the case of precipitation of meteoric water, we can calculate the isotopic ratios for the two phases at any given time, while considering an open system, thus being in purely Rayleigh distillation conditions. Using the derivation of the initial Rayleigh distillation equation above, we can define the oxygen isotope ration for the forming condensate as

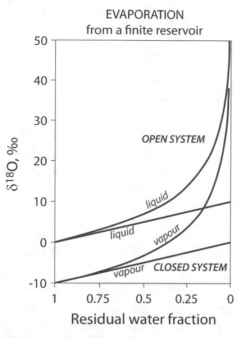

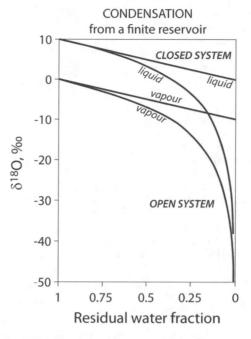

Fig. 3.4 Evolution of the oxygen isotopic composition for closed and open systems regarding evaporation and condensation from a finite reservoir (i.e., a closed basin such as lake or sea, or a single air mass). Notice that if the system is closed (blue lines), with the two phases in contact and in isotopic equilibrium, the final product, when all water is either fully evaporated or fully precipitated, is isotopically equivalent to the initial source. However, this is not the case in an open system (Rayleigh distillation, red lines), where the product is removed from the system and is not in isotopic equilibrium with the initial source: this situation results in constantly increasing (in the case of evaporation) or decreasing (in the case of condensation) isotopic ratios

$$\frac{\left(^{18}O/^{16}O\right)_{condensate}}{\left(^{18}O/^{16}O\right)_{initial\ vapour}} = \alpha f^{\alpha-1}$$

And, for the remaining vapour:

$$\frac{\left(^{18}O/^{16}O\right)_{remaning\ vapour}}{\left(^{18}O/^{16}O\right)_{initial\ vapour}} = f^{\alpha-1}$$

These equations can be written for the delta values, as well, for both oxygen and hydrogen, and the results can be graphically presented, as shown in Fig. 3.4.

As seen, the hydrogen and oxygen isotopic compositions of the global meteoric waters correspond to Rayleigh distillation and can be calculated as a function of vapour remaining. However, an additional complication arises from the fact that, as the air moisture travels away from the equator and closer to the poles, air temperatures decrease and the fractionation factor increases (Fig. 3.5). This is expressed by the following equation:

$$\frac{\left(^{18}O/^{16}O\right)_{condensate}}{\left(^{18}O/^{16}O\right)_{initial\ vapour}} = \frac{\alpha}{\alpha_{initial}} f^{\alpha-1}$$

where α is the fractionation factor for the temperature at which each specific precipitation occurs. The temperature effect can be visualized in the diagram relating Rayleigh distillation and global water cycle (Fig. 3.5).

3.5   Effects of Evaporation

As we noted earlier, kinetic fractionation occurs during evaporation of ocean water, as the vapour is removed from contact with the liquid and they are thus not in equilibrium. Let us consider a body of water undergoing a certain degree of evaporation. As the two main constituents of water—hydrogen and oxygen—have very different effective ionic radii (hydrogen is much smaller than oxygen), their kinetic fractionation differs significantly. Thus, fractionation during evaporation will not follow the "8-times-stronger-for-hydrogen" rule, which is based on the differences in mass. Rather, hydrogen isotopes will fractionate typically 3–6 times (approximately 4.5 on average) more than oxygen isotopes, based on empirical observations. The specific slope for each evaporative process will depend predominantly on the humidity of the air into which the water evaporates: more humid environments will produce fractionations closer to that produced at equilibrium as there is more humidity in the air with which the evaporated water can stay in (partial)

Fig. 3.5 Rayleigh distillation and the global water precipitation, underlining the temperature effect: as the air mass moves away from the intertropical regions and towards the poles, precipitation will occur at lower temperatures, increasing the fractionation between vapour and liquid. Notice the jump of fractionation factor between vapour and solid (snow), due to the stronger bonding between atoms in the crystal structure of ice relative to water. This leads to increased preferential incorporation ot ^{18}O and ^{2}H in snow relative to liquid water. After Clark and Fritz [1], modified

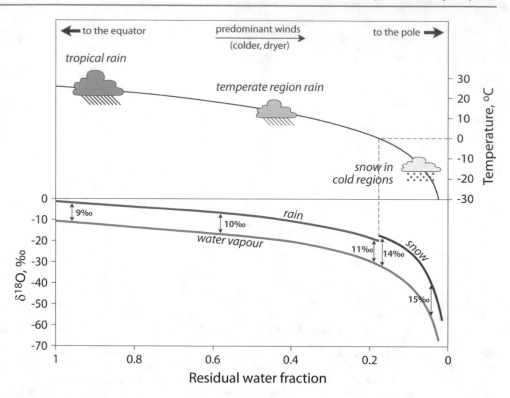

equilibrium. On the opposite end, more arid environments will produce more kinetic fractionation and lower the slope of the evaporation line. Thus, water that has been affected by any degree of evaporation will lie on a distinct line in the $\delta^{18}O$ versus $\delta^{2}H$ diagram (Fig. 3.6).

For each evaporation process, two distinct lines will form, parallel but in opposite direction: one is for the remaining liquid, and the other for the vapour (we rarely have access to the vapour phase, but can observe the liquid phase). The two lines will originate from the same position on the meteoric water line corresponding to the geographic location of the river.

A very different trend is observed when saline water is evaporated from a closed basin. Initially, the isotopic ratios of the water will increase for both hydrogen and oxygen, following a straight line. (Incidentally, this is the only way to produce water with $\delta^{18}O$ and $\delta^{2}H$ values above 0 in the natural environment.) However, as water salinity increases beyond a certain point, several chloride and sulphate minerals (e.g., carnalite, $KMgCl_3 \ast 6H_2O$, and gypsum, $CaSO_4 \ast 2H_2O$) will start to precipitate, some of which incorporate hydrogen, or oxygen, or both. As atoms are bound much stronger in crystals than in water, the minerals will tend to preferentially incorporate the heavy isotope, leaving the light isotope in the water. As this process accelerates, the direction of the evaporation line will tend to curve and eventually completely inverse direction: this happens when fractionation from precipitation of minerals

(which occurs at equilibrium) overtakes fractionation from evaporation.

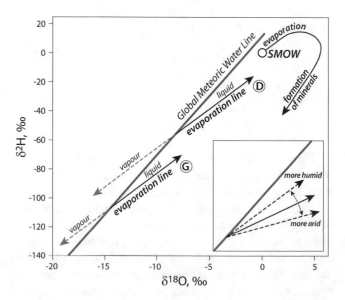

Fig. 3.6 Effects of evaporation from freshwater sources or seawater, using two examples: Rio Grande in the USA and Mexico (G) and the river Darling in Australia (D), both with a slope (the extent of fractionation for hydrogen vs. the extent of fractionation for oxygen) of approximately 5. The humidity of the air is the most significant effect on the slope of the evaporation line. Evaporation from a closed sea basin produces a particular line due to formation *of minerals*, as described in the text

3.6 Isotope Mixture

So far, we have neglected to consider one major part of the global water cycle, the return of meteoric water back into the ocean. This occurs mostly through direct precipitation over the ocean (approximately 90%), with the rest being surface and ground flow. This brings us to the very important concept of isotopic mixture. In order to approach this notion, we will imagine two rivers, each with their own hydrogen and oxygen isotopic composition, meeting and forming a new river (Fig. 3.7).

Can we calculate the isotopic composition of the new river, based on the compositions of the two initial ones? The calculation is actually a simple one and, more importantly, will always hold true and can be used for any isotopic mixture calculations. For the example of oxygen, the isotopic mixture equation is:

$$\delta^{18}O_C = \delta^{18}O_A f_A \left([O]_A/[O]_C\right) + \delta^{18}O_B (1 - f_A) \left([O]_B/[O]_C\right)$$

where f_A is the fraction of volume of River A, calculated by

$$f_A = V_A/(V_A + V_B) = V_A/V_C$$

(Also, $f_A + f_B = 1$, hence $f_B = 1 - f_A$)

In other words, calculating the isotopic composition of a binary mixture takes into account the concentration of the element of interest (oxygen in this example) in each initial part, the proportion of volume of each part, and of course, the isotopic composition of each part. Let us reiterate this, as it is important: to calculate the isotopic composition of a binary mixture, we must use the isotopic composition of the two compounds and two "weighing factors," which are based on the volume proportion of each source (f_A and f_B) and the concentration of the element of interest in each source ($[O]_A$ and $[O]_B$ in this example).

Using simple mathematics, we can rewrite the isotopic mixture equation to deduce the composition of one of the initial sources, by knowing the isotopic composition of the other source and of the mixture. We can also calculate the proportions of each source if we know the isotopic compositions of the two sources and of the mixture. Possibilities are endless and these calculations are very useful, therefore let us remember this simple calculation as it will serve us well. We can also use isotope ratios rather than the delta values since, mathematically, it comes to exactly the same. Finally, we can develop this calculation for an isotopic mixture of more than two compounds, which will be very helpful, as we will see in Chap. 5.

In the case of river water, things are quite easy, as the concentration of oxygen remains the same: $[O]_A = [O]_B$, simplifying the equation to:

$$\delta^{18}O_C = \delta^{18}O_A f_A + \delta^{18}O_B (1 - f_A)$$

In other words, only the volume proportion of each source (the weighing factor) and their respective isotopic compositions will be used to calculate the mixture isotopic composition, making things quite simple.

The calculation of an isotopic mixture when the concentrations of the element of interest are different can be best approached visually (Fig. 3.8).

Finally, we can complicate things a bit more: how do we calculate the isotopic mixture between two or more sources, for isotopes of two different elements? Simply put, we will use the exact same logic as for one element, and the corresponding derivations of the main equation: we will take into account the proportions of the different sources, the concentrations of the elements of interest in each source, and the

Fig. 3.7 Two rivers, A and B, meeting to form a new one, C: how can we calculate the oxygen isotopic composition of the new river, $\delta^{18}O_C$?

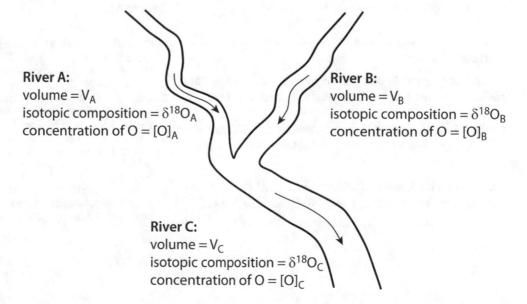

River A:
volume $= V_A$
isotopic composition $= \delta^{18}O_A$
concentration of O $= [O]_A$

River B:
volume $= V_B$
isotopic composition $= \delta^{18}O_B$
concentration of O $= [O]_B$

River C:
volume $= V_C$
isotopic composition $= \delta^{18}O_C$
concentration of O $= [O]_C$

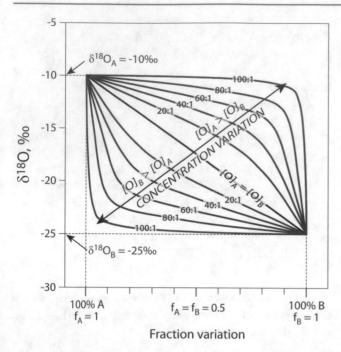

Fig. 3.8 Graphic representation of the isotopic mixture calculation, for the same element, of two sources: we must take into account the concentrations of the element in the two sources, the proportion of each source, and their isotopic compositions. Each mixture will be found on a line corresponding to the concentration proportion of the two sources and the exact position will depend on the volume fraction of each source. This illustration demonstrates the significant influence of element concentration on the isotopic composition of the mixture

isotopic compositions for the two different elements. As concentrations of the two elements in the different sources will often vary, the graphic representation of such a mixture calculation will closely resemble that in Fig. 3.8 (when only two sources are considered and with isotopic compositions of the two different elements placed on the two axes). When the mixture of more than two sources or compounds is considered, the calculations become a bit more involved (although, as mentioned, the logic remains the same) and sometimes necessitate computer-based modelling.

Further Reading

Several very good books, or chapters in other books, are published on the subject, including this short selection:

Isotope Tracers in Catchment Hydrology, C. Kendall and J.J. McDonnell, Editors, Elsevier, 1998, ISBN 978-008-092915-6.

Environmental Isotopes in Hydrogeology, I.D. Clark and P. Fritz, Lewis Publishers, 1997, ISBN 1-56670-249-6.

Stable Isotope Geochemistry: a Tribute to Samuel Epstein, H.P. Taylor, J.R. O'Neil, and I.R. Kaplan, Editors. The Geochemical Society Special Publication #3, ISBN 0-941809-02-1. Of particular interest here is Part B, The Hydrosphere and Ancient Oceans.

Isotopes in the Water Cycle: Past, Present and Future of a Developing Science, P.K. Aggarwal, J.R. Gat, and K.F. Froehlich, Editors, Springer Science, 2006, ISBN 978-140-203023-9.

Isotope Hydrology, P.K. Aggarwal, Editor, IAHS Press, 2012, ISBN 978-190-716129-2.

Environmental Isotopes in the Hydrological Cycle: Principles and Applications, W.G. Mook. IHP-V Technical Documents in Hydrology, N° 39. UNESCO—IAEA 2001.

Questions

- What are the main reservoirs of water on Earth?
- What sort of fractionation occurs during evaporation and condensation?
- What is the relationship between the extent of fractionation between hydrogen and oxygen isotopes during the water cycle? What is it defined by? What is this relationship called?
- What factors, and to what extent, affect the meteoric water isotopes composition?
- What is Rayleigh distillation? How does it work?
- What are the effects of evaporation, and how will they differ according to changing environmental conditions?
- What is the basic isotopic mixture equation, and how can it be modified to suit our needs? What is its graphic representation?

References

1. Clark ID, Fritz P (1997) Environmental isotopes in hydrogeology, Lewis Publishers, New York, 328 p
2. Dansgaard W (1964) Stable isotopes in precipitations. Tellus 16:436–463
3. Gat JR (1980) The isotopes of hydrogen and oxygen in precipitations. In: Fritz P, Fontes JC (eds) Handbook of environmental isotope geochemistry, vol 1. Elsevier Scientific, Amsterdam, pp 21–47
4. Terzer S, Wassenaar LI, Araguás-Araguás LJ, Aggarwal PK (2013) Global isoscapes for ^{18}O and ^{2}H in precipitation: improved prediction using regionalized climatic regression models. Hydrol Earth Syst Sci 17:4713–4728
5. Yurtsever Y (1975) Worldwide survey of stable isotopes in precipitations. Report of the Section on Isotope Hydrology, International Atomic Agency, Vienna

Geology (the study of the Earth and the rocks and minerals within it) is the one science that most heavily relies on and is assisted by the use of isotopes as an investigation tool. The recent period of strong and sustained development in geological understanding—starting in the 1950s—coincides with very significant advances in isotope research (and in geophysics), itself strongly dependent on technological advances such as electronics and modern computers. The reason for this quasi-dependence of geology on isotopes (to the point of colloquially equating geochemistry to *isotope* geochemistry), is that every and each process that occurred in the past and affected a rock and its components also affected the isotopic composition of many elements within the rock. By studying the isotopic composition of a geological sample, we are able to understand not only what processes affected that rock, but also when and under what physical and chemical conditions these processes occurred. As a consequence, isotope geochemistry is very well understood and widely applied. In this chapter, we will briefly examine some of the many applications of isotopes, both stable and radiogenic, in geology.

4.1 Stable Isotopes and Fluids Interacting with Rocks

Fractionation

One particularity of geological processes is their relatively low speed, compared to those in other parts of the Earth, such as the biosphere and the hydrosphere, for instance: the timescale in geology is typically measured in millions of years, and in other domains, it is measured in days to years. One result is that the fractionation affecting stable isotopes during geological processes is almost always under equilibrium conditions: the different components in any reaction will have been in contact sufficiently long enough to be in isotopic equilibrium. The first and most beneficial consequence is that fractionation factors—albeit low—can be known with a good level of precision. The relationship between fractionation factors and temperature is well established for a great variety of reactions commonly occurring in geology, based on empirical observations and laboratory experiments. We will soon see the importance of these considerations.

Water is a major participant in the many reactions that occur involving rocks. Since water is abundant, it is in constant interaction with minerals and rocks and can dissolve them and mobilize their chemical elements; water can alter rocks and minerals by forming new minerals, and minerals can also precipitate from water. Let us consider the example of quartz precipitating from water, which is a very common reaction (Fig. 4.1):

$$H_4SiO_4 \rightarrow SiO_2 + 2H_2O$$

If water's oxygen isotopic composition is -15‰ and quartz's oxygen isotopic composition is -10‰, the difference of 5‰ will let us know the temperature at which the reaction took place: we can calculate this temperature according to the equation that relates fractionation factor and temperature (established experimentally by Clayton et al. [3]; one amongst many researchers on the topic):

$$\delta^{18}O_{quartz} - \delta^{18}O_{water} \sim 1000\ln \alpha = 3.38\left(10^6/T^2\right) - 3.34$$

where T is the reaction temperature in degrees Kelvin (K). (The inverse relationship between temperature and extent of equilibrium fractionation, described in Chap. 1, is clearly apparent in this equation.) We can calculate that the temperature at which the reaction took place was approximately 736 K, or approximately 363 °C.

However, in a geological sample taken today, we will have access only to one part of this reaction, namely its product, quartz. In order to calculate the isotopic composition of the water from which the quartz precipitated (and which is long gone), we must have some independent measure or at least have a fairly reliable estimate, of the

© Springer Nature Switzerland AG 2020
P. Alexandre, *Isotopes and the Natural Environment*, Springer Textbooks in Earth Sciences,
Geography and Environment, https://doi.org/10.1007/978-3-030-33652-3_4

Fig. 4.1 Equilibrium fractionation of oxygen isotopes during precipitation of quartz from water. By knowing the difference in isotopic composition between water and quarts, the reaction temperature can be reliably calculated

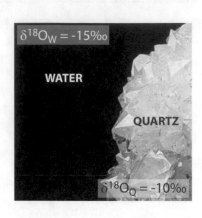

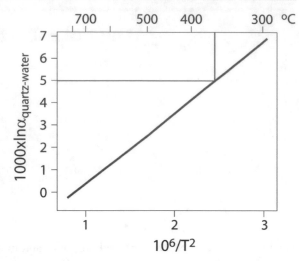

precipitation temperature. This is often challenging, but there are several methods that allow us to estimate the temperatures of past geological events and specific reactions.

If we have two mineral phases, e.g., quartz and calcite, that precipitated at the same time and under equilibrium (which we can often ascertain by microscopic observations), we can use the fractionation factor between them, or the combination of the fractionation factors between them and the water (Fig. 4.2), to calculate the temperature at which they precipitated. This is a very significant and important use of stable isotopes, as it provides a reliable and precise measure of paleo-temperatures. It can be applied in a variety

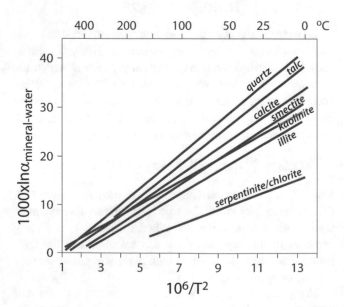

Fig. 4.2 Fractionation of oxygen isotopes in the most common mineral water systems at the surface of the Earth. In general, the extent of fractionation, $\delta^{18}O_{mineral}-\delta^{18}O_{water}$, is linearly inversely dependent on T ($\ln\alpha_{mineral\ water} \sim a + b/T$), for lower temperatures, and on T^2, at higher temperatures ($\ln\alpha_{mineral\ water} \sim a + b/T^2$). Data from Bottinga and Javoy [1] and Kyser [6]

of situations when two mineral phases are formed in equilibrium, as the extent of fractionation as a function of temperature, between most common minerals, is well established.

Water–rock interactions

Most rocks contain a certain amount of water, which can be sometimes minute (in igneous and high-grade metamorphic rocks) and sometimes significant (e.g., in sedimentary rocks). Significantly, these waters will have specific isotopic compositions (Fig. 4.3) that allow us to appreciate the extent of water–rock interactions.

Let us imagine a situation regarding meteoric water, its hydrogen and oxygen isotopic composition lying somewhere on the meteoric water line, reacting with magmatic or metamorphic rocks. Isotopically, the first significant change that will affect this water, at high water–rock ratios, will modify its oxygen isotopes, as oxygen is a major component in all rocks, at approximately 50 weight percent (wt%), whereas hydrogen is not (much less than a percent). The outcome is moving the water's oxygen isotopic composition towards higher values, but not changing in any significant way its hydrogen isotopic composition. We will call these waters "^{18}O-shifted" waters (Fig. 4.3): their oxygen is being equilibrated, to a significant extent, with that of the rocks, but the hydrogen is not and remains very similar to that of the original meteoric water. This situation is very commonly observed in geothermal waters, such as in the Yellowstone National Park in the USA (Fig. 4.3). If waters penetrate further into the rocks, the water/rock ratio will decrease to a point where the hydrogen in the rocks (contained in hydrated minerals such as micas and amphibole) will start to isotopically equilibrate with the hydrogen in the waters. This exchange results in increased δ^2H water values, eventually water fully isotopically equilibrating with the rock (Fig. 4.3).

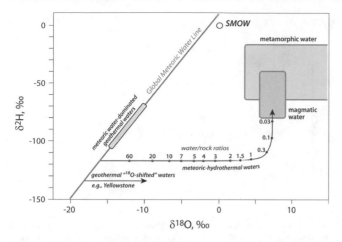

Fig. 4.3 Hydrogen and oxygen isotopic compositions of the principal terrestrial water reservoirs, including the approximate fields corresponding to waters contained in magmatic and metamorphic rocks. The field of sedimentary rocks (not represented in the diagram) is less well defined and more variable. Horizontal lines represent waters that have been affected by high water/rock ratio interaction, which is typical of geothermal fields such as Yellowstone in the USA. For very low water/rock ratios, hydrogen isotopes will also be affected. The isotopic composition of the water can be calculated following the general mixing rules stated in Chap. 3 (Fig. 3.8)

This isotopic exchange follows the general rules of isotopic mixture described in the previous chapter and specifically the situation where the two mixing sources have very different concentrations of the elements of interest, O and H in this case (Fig. 3.8).

Because of the very different concentrations of oxygen and hydrogen in water, compared to rocks, the mixing lines produced by water–rock interactions at a wide range of water/rock ratios are never straight lines (Fig. 4.3). The only straight lines in the δ^2H versus $\delta^{18}O$ diagram are the meteoric water line and the evaporations lines that were described in the previous chapter (Fig. 3.6).

Based on the above observations, we can derive the origin of any specific water sample that interacted with a rock to some degree. For water/rock ratios above approximately 1, it is possible to extrapolate as to where the mixing line intercepts the meteoric water line and thus obtain the initial hydrogen and oxygen isotopic compositions of the surface water that interacted with the rocks.

A word of caution regarding the use of the δ^2H versus $\delta^{18}O$ diagram: this is a diagram used uniquely for waters, with specific and exact locations of the major Earth water reservoirs. This means that every time we obtain the hydrogen and oxygen isotopic composition for a specific mineral, we must calculate the isotopic compositions for the water that formed and was *in equilibrium* with that mineral (we will assume

equilibrium, unless we have clear indication of the contrary), using the formation temperature (obtained by an *independent* method) and the appropriate equation for that mineral. If we do not have access to the temperature and not even an educated guess about it, we cannot use the mineral data in this diagram: again, this diagram is only to be used for plotting and comparing waters.

Ore deposits

One very useful application of this approach (the examination the isotopic effects of the water–rock interactions) is the study of ore deposits: the major vector, or transportation medium, that leads to the formation of the vast majority of deposits is water-dominated fluid. This fluid often contains a certain amount of dissolved salts (anywhere up to 30%), such as NaCl, $CaCl_2$, or $MgCl_2$. In general terms, the fluid mobilises the element of interest (or commodity element) from its source and transports it to the appropriate location of precipitation, where a deposit forms within a physical or chemical "trap", or a combination of both. During the deposit-forming process, the fluid interacts with the host rock, which leads to a complex series of alterations and a set of alteration-produced minerals. Each deposit and deposit type will have its own cortege of alteration noneconomic minerals, with the most common being clay minerals (e.g., illite/sericite, kaolinite, chlorite),

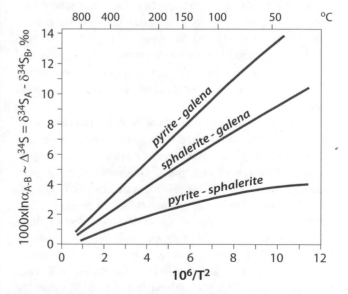

Fig. 4.4 Sulphur isotopes fractionation factors as a function of temperature between pyrite, sphalerite, and galena, which are common sulphide minerals in many zinc and lead deposits and deposit types. Data from Kyser [6]

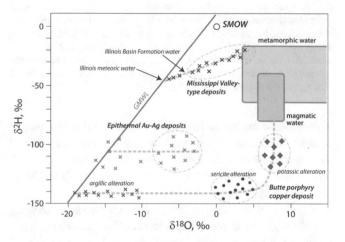

Fig. 4.5 Example of sulphur isotope measurements applied to the study of ore deposits. The samples come from a base metals and gold volcanogenic massive sulphide deposit in northern Manitoba, Canada (author's own research). When the Δ^{34}S between the three possible pairs of minerals, pyrite–sphalerite (Py-Sp), pyrite–galena (Py-Gn), and sphalerite–galena (Sp-Gn), are considered, a temperature of about 500 °C can be derived, based on the fractionation factors given in Fig. 4.4

quartz, carbonates (calcite, ankerite, dolomite), and sulphides (pyrite, pyrrhotite).

If we have two of these minerals forming at the same time —and thus at isotopic equilibrium—as confirmed by microscopic observations (e.g., quartz and calcite, for vein gold deposits, or pyrite–pyrrhotite for a MVT deposit), we can derive their formation temperature based on the fractionation equation for the two specific minerals (Figs. 4.4, 4.5). On the other hand, if we have access to the temperature from a couple of co-forming minerals—or through an *independent* (non-isotopic) method—we can derive the isotopic composition of the water that was in equilibrium with a specific mineral (for instance, sericite or chlorite). This allows us to compare the isotopic composition of this water and ascertain its origin: was it ocean water, meteoric water, or maybe evaporated water (Fig. 4.6)? Knowing the deposit formation temperature and the source of the water involved in the ore deposition helps us enormously to decipher and understand the formation of any deposit: indeed, stable isotopes are a major investigation tool for ore deposit geology.

Finally, stable isotopes can be an actual—and effective— exploration tool: both O and S isotopes have been demonstrated to show clear variation as function of proximity to ore or as function of ore grade. One such case has been documented for the volcanogenic massive sulphide deposits of the Flin Flon-Snow Lake Mineral Belt in northern Manitoba, Canada, where several Zn-Cu-Pb(+Au) deposits have been discovered [7]. Sulphide minerals from several ore bodies and from barren areas were extracted and analysed for sulphur isotopes and the results show clear difference between the two (Fig. 4.7). Above a certain δ^{34}S value, the probability of finding a mineralized ore body decreased sharply. This is possibly explained by the change of sulphur isotopic composition as a function of sulphur's oxidation

Fig. 4.6 Isotopic composition of ore-forming fluids for three different types of deposits from the USA: the epithermal (low temperature) Au-Ag deposits, the Mississippi Valley-type deposits in the Illinois, and the Butte porphyry copper deposit in Montana. Each type of deposit was formed by a different variety of water: the Mississippi Valley-type deposits by the strongly evaporated meteoric water in the Illinois Basin, the epithermal Au-Ag deposits by higher latitude meteoric water that had undergone weak interaction with the host rocks (high water/rock ratio), and the Butte deposit by strongly evolved meteoric water from even higher latitudes. In this last case, the progression of alteration, from low-temperature argillic to the high-temperature potassic alteration, can be followed by the progressively lower water/rock ratio and eventual isotopic equilibration with the host rock, the Butte quartz monzonite. Data from Kyser [6]

state: it is consistently higher in oxidized sulphur (as in sulphates) than in reduced one (as in sulphides), by typically about 10 to 20‰. Other examples of stable isotopes applied to exploration include carbon isotopes in oil exploration, where higher δ^{13}C values of hydrocarbons detected at surface correspond to higher level of hydrocarbon maturation, or oxygen isotopes in gold exploration, where lower δ^{18}O have been observed to correlate well with the presence of productive gold deposits.

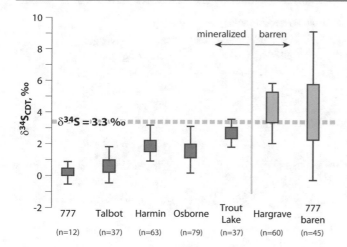

Fig. 4.7 Sulphur isotopic composition of sulphide minerals from mineralized and barren systems in northern Manitoba, Canada. The chart shows a clear distinction between the two types of systems, possibly due to sulphur's oxidation state, itself due to the prevailing Eh conditions during deposit formation [7]

4.2   Geochronology and the Timing of Events

In Chap. 1, we discussed that radioactive isotopes disintegrate at a constant speed, meaning that if we know that speed and the amounts of parent and daughter isotopes in a geological sample (mineral or rock), we can calculate its age, by using the equation there. We can rewrite this equation for the ^{235}U—^{207}Pb parent–daughter couple, for instance:

$$Age_{207Pb/235U} = \frac{1}{\lambda_{235U}} \left(^{207}Pb/^{235}U + 1 \right)$$

We can also write a very similar equation for the other uranium isotope, ^{208}U, which disintegrates to ^{206}Pb, giving us two ages for the uranium–lead series. More on this later.

While the general principle of absolute (or isotope) geochronology is simple, the application of the method and the interpretations involved can be rather complex. Firstly, there may be several radioactive isotopes incorporated into a mineral during its formation. Of these, the most common are ^{40}K, ^{87}Rb, ^{147}Sm, ^{176}Lu, ^{235}U, ^{238}U, and ^{232}Th (Table 4.1). In practical terms, only one isotopic system is used when any particular mineral is dated, but in some cases, we can use more than one isotopic system.

Another complication may arise from the way the daughter isotope is accumulated into the mineral analysed. Whereas the parent isotope is most commonly chemically bound within the crystalline structure of the mineral, this is not the case for the daughter isotope, which is often mechanically retained but without being chemically bound. As a result, the daughter isotope will tend to leave the mineral, most commonly by diffusion. The main factor,

among several, controlling the diffusion of a chemical element in a crystalline structure is temperature: the higher the temperature, the higher the diffusion rate and the faster the daughter element will leave. Let us imagine a mineral that has formed at a high temperature (perhaps in a magma chamber), such that most of the daughter isotope has left the system by diffusion. As the rock containing the mineral cools down, the diffusion rate will become lower and lower, to the point where most—and then all—of the daughter isotope will be retained and will start accumulating within the mineral. This is a critical point for the understanding of isotope geochronology: when we analyse a mineral and calculate an isotopic age for it, that age corresponds to the time when the mineral cooled sufficiently enough to retain the all of the daughter isotope, and not to the time when the mineral formed (Fig. 4.8). Indeed, there are many cases when the mineral formed several millions of years before the system became closed and the daughter isotopes started to cumulate.

The temperature below which the entirety of the daughter isotope is retained is called closure temperature (or blocking temperature) and is different for each mineral and for each isotopic system (Table 4.2). This means that if we date a mineral using two different isotopic systems, we will find two different ages. These ages may be very different, and one of the two may be very close in time to the formation of the mineral, therefore we must be very careful when we interpret the data.

There are two specific implications that arise from the concept of closure temperature. Firstly, if a mineral was formed at temperatures close to or lower than the closure

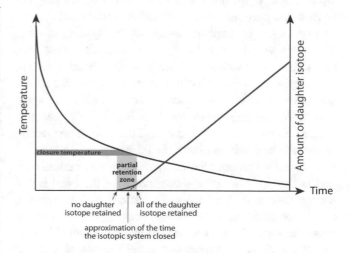

Fig. 4.8 Graphic representation of the notion of closure temperature and the corresponding isotopic age. The isotopic system will become closed only below a certain temperature, and is different for each isotope and each mineral. The time that elapsed before the system became closed cannot be found on the basis of a single geochronological method

Table 4.1 The parent–daughter couples most commonly used in absolute (isotope) geochronology

Parent isotope	Daughter isotope	Main disintegration path	Parent isotope abundance (%)	Half-life (years)	Disintegration constant $\lambda(y^{-1})$	Minerals typically incorporating the parent isotope
^{40}K	^{40}Ar	β^-, ec	0.001167	1.19×10^{10}	5.5492×10^{-10}	Any K-rick mineral
^{87}Rb	^{87}Sr	β^-	27.8346	48.8×10^9	1.42×10^{-11}	Any K-rick mineral
^{147}Sm	^{143}Nd	α	15.0	1.06×10^{11}	6.54×10^{-12}	Various
^{176}Lu	^{176}Hf	β^-	2.6	3.6×10^{10}	1.867×10^{-11}	Garnet, phosphates
^{235}U	^{207}Pb	α	99.2743	7.07×10^8	9.8571×10^{-10}	Zircon, monazite, apatite
^{238}U	^{206}Pb	α	0.7200	4.47×10^9	1.55125×10^{-10}	Zircon, monazite, apatite
^{232}Th	^{208}Pb	α	100.0	1.4×10^{10}	4.948×10^{-11}	Zircon, monazite, apatite

Data from Brownlow [2]

temperatures for the particular isotopic system used for dating, then the age obtained will be that of the mineral formation. For instance, if a zircon formed at 850 °C, then the age obtained by U-Pb dating will be identical to the crystallization age for that zircon, as its closure temperature for Pb is higher. As another example, if a muscovite formed at 300 °C, the K-Ar age obtained will be the crystallization age for this mineral, as the Ar closure temperature for muscovite is approximately 350 °C. Thus, we can have access to the actual formation of a mineral and a rock, on condition that we select the appropriate isotopic system and the appropriate mineral.

On the other hand, what would happen if a mineral was heated, at some point during its history, at temperatures above the closure temperatures for the specific isotopic system? It would simply lose some or all of the radiogenic daughter isotope, partially or completely resetting the isotopic "clock" of the mineral. If all of the radiogenic daughter isotope had been lost, then the age we would obtain would be that of the subsequent closure of the system; in other words, we would obtain the age of the reheating event. This has some disadvantages, as we lose the initial age, but, encouragingly, we are able to obtain the age of the subsequent event. The worst-case scenario occurs when only some portion of the radiogenic daughter isotope has been lost during a thermal event: it is difficult, though not impossible, as we will later see, to obtain the initial age.

The major consequence of the closure temperature concept is that we must always consider what the event is that we are dating: such as initial formation, cooling below a certain temperature, or reheating at some point after initial system closure. We must remember that we do not obtain the age of a rock or a mineral, but rather the age of an event affecting that rock or mineral. This distinction is important and it is good policy to keep it in mind and use it appropriately.

Another significant complication that arises when applying the method is the fact that the mineral could incorporate, during its formation, some amount of the radiogenic daughter isotope. In some cases, we have good grounds to assume that no daughter isotope was initially present (for instance, when there was very little of it or when the mineral could not incorporate it into its crystalline structure), but often we suspect that at least some small amount is present. In other words, the daughter isotope that we measure may not be only the product of radioactive disintegration of the parent isotope, but may also be originally present in the mineral, as in the instance of the U-Pb series:

$$^{207}Pb_{measured\,now} = {}^{207}Pb_{radiogenic} + {}^{207}Pb_{original}$$

There are two methods to calculate and subtract the original amount of daughter isotope: the common Pb correction, and the isochron method.

The common lead correction relies on the existence of ^{204}Pb, or common lead, which is a lead isotope that is stable and non-radiogenic: its amount on Earth is constant and is the same as it was when the Earth initially formed (this is why we also call it *primordial*). Crucially, some parts of the radiogenic ^{207}Pb and ^{206}Pb were also initially present, but their amounts grew over time as more was produced by disintegration of ^{235}U and ^{238}U, respectively. At the moment when the mineral (for instance, zircon) formed, it incorporated some amounts of ^{204}Pb, ^{207}Pb, and ^{235}U (we will consider only one of the two U series, but it will work in exactly the same way as for the other). Most importantly, the ratio $^{207}Pb/^{204}Pb$ for that time is known, as radiogenic lead/common lead ratios are well known for the entire history of the Earth. By using the $^{207}Pb/^{204}Pb$ ratio for the mineral formation time, we can account for the amount of ^{207}Pb that was incorporated into the mineral at that time:

$$^{207}Pb_{radiogenic} = {}^{207}Pb_{measured\,now} - {}^{207}Pb_{original}$$
$$= {}^{207}Pb_{measured\,now} - R_t{}^{204}Pb$$

where $R_t = {}^{207}Pb/^{204}Pb$ at the time the mineral formed.

The main difficulty here is that we do not initially know the formation time of the mineral, so we must use an iterative approach: we start by making the best possible guess of the age, calculate an age and then recalculate a second or a third age, each time using the age from the previous iteration.

The isochrone method also relies on the existence of ^{204}Pb. We will use the general age equation

$$^{207}Pbb_{now} = {}^{207}Pb_{original} + {}^{235}U_{now}(e^{\lambda t-1})$$

and then divide each isotope in it by ^{204}Pb:

$$\left(^{207}Pb/^{204}Pb\right)_{now} = \left(^{207}Pb/^{204}Pb\right)_{original} + \left(^{235}U/^{204}Pb\right)_{now}(e^{\lambda t} - 1)$$

In this new equation, only the $\left(^{207}Pb/^{204}Pb\right)_{original}$ is not known, whereas the other two ratios are measured. We will also notice that this is a linear equation: y = a +bx, where

y $\left(^{207}Pb/^{204}Pb\right)_{now}$, which is measured
x $\left(^{235}U/^{204}Pb\right)_{now}$, which is also measured
a $\left(^{207}Pb/^{204}Pb\right)_{original}$, which is the intercept of the line at x = 0, and
b $(e^{\lambda t}-1)$, which is the slope of the line (Fig. 4.9).

This method relies on the analysis of several minerals that were formed at the same time, but incorporated different amounts of uranium. Once the slope of the isochron ("line of the same age") is obtained, we can calculate the age of the rock of which the minerals analysed are part.

This is a very powerful and reliable method and is very commonly used, not only for the U-Pb systems, but also for the ^{87}Rb-^{87}Sr, ^{147}Sm-^{143}Nd, and ^{176}Lu-^{176}Hf ones and several others. We can rewrite the isochron equation for ^{176}Lu-^{176}Hf system, for instance, by using the stable and non-radiogenic ^{177}Hf:

$$\left(^{176}Hf/^{177}Hf\right)_{now} = \left(^{176}Hf/^{177}Hf\right)_{original} + \left(^{176}Lu/^{177}H\right)_{now}(e^{\lambda t} - 1)$$

Let us now return to the U-Pb system, which has two parent–daughter pairs, ^{235}U-^{207}Pb and ^{238}U-^{206}Pb. Once the common Pb has been subtracted, the two age equations can be written

$$Age_{^{207}Pb/^{235}U} = \frac{1}{\lambda_{235}U}\left(^{207}Pb^*/^{235}U + 1\right)$$

and

$$Age_{^{206}Pb/^{238}U} = \frac{1}{\lambda_{238}U}\left(^{206}Pb^*/^{238}U + 1\right)$$

where * denotes lead generated only by the disintegration of U, or radiogenic lead.

The fact that two ages can be derived from the U-Pb system is unique and gives us a powerful tool for deciphering the history of a sample. This tool is called the Concordia diagram (Fig. 4.10). The evolution of the terrestrial $^{207}Pb/^{235}U$ and $^{206}Pb/^{238}U$ ratios are known, as they are

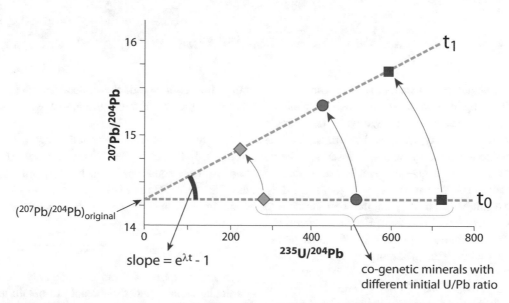

Fig. 4.9 U-Pb isochron diagram. Several different minerals (or whole rock samples) with different initial U/Pb ratios are used. As the ^{235}U progressively disintegrates to ^{207}Pb, the $^{235}U/^{204}Pb$ ratio will decrease and the $^{207}P/^{204}Pb$ ratio will increase. This results in a higher slope of the isochron line, corresponding to a higher age. The initial $^{207}P/^{204}Pb$ ratio remains the same, of course, and can be derived from the intercept of the isochron at $^{235}U/^{204}Pb = 0$. The minerals' isotopic composition will always be on the same line

Fig. 4.10 Concordia diagram, using both $^{206}Pb*/^{238}U$ and $^{207}Pb*/^{235}U$ ratios. The Concordia line is defined as the line on which the two U-Pb ages are the same, whereas the Discordia line is situated between two ages, namely the initial crystallizations age, at the upper intercept, and the perturbation age leading to partial Pb* loss, at the lower intercept

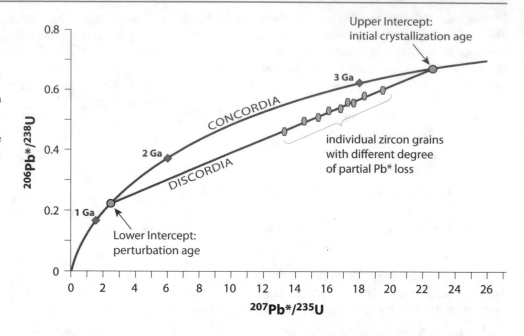

Table 4.2 Typical closure temperatures for different minerals and different parent–daughter systems. The exact and specific closure temperatures can be sometimes difficult to assess, as they are fairly variable and depend on many factors, including grain size and chemical composition of the mineral, the cooling rate, and possibly pressure

System	Mineral	Approximate closure temperature (°C)
^{40}K-^{40}Ar	Hornblende	550
	Muscovite	350
	Biotite	300
	Feldspar	200
^{87}Rb-^{87}Sr	Muscovite	450
	Biotite	350
^{235}U-^{207}Pb and ^{238}U-^{206}Pb	Zircon	900
	Monazite	800
	Titanite	600
	Apatite	450
	Rutile	400

based on the disintegration constants given in Table 4.1. If the two U-Pb ages obtained on the same mineral, typically a zircon, are the same, then this is the age when the isotopic system had initially closed (typically at the formation of the mineral), *or* the age when the system was closed after being completely reset during a reheating event. However, the Pb closure temperature for zircon is fairly high (Table 4.2), meaning that we rarely reach temperatures that are able to completely reset zircon's U-Pb system. However, zircon is a very resistant mineral and a single crystal can often participate in numerous igneous or metamorphic events, or both, so that its isotopic system will be affected by partial Pb* loss. In that case, the obtained $^{206}Pb*/^{238}U$ and $^{207}Pb*/^{235}U$ ratios will not lie on the Concordia line, but on a straight line, called Discordia, that connects two Concordia/Discordia intercepts (Fig. 4.10). The most common interpretation of such a situation is that the upper intercept represents the

initial formation age of the mineral and the lower intercept, the age of the perturbation (thermal) event causing the partial Pb* loss.

Finally, we can consider another age calculation issuing from the U-Pb series: the $^{207}Pb/^{206}Pb$ age. If we divide the two U-Pb age equations by each other, and take into account that the $^{238}U/^{235}U$ ratio is known and constant at 137.88, we come up with the $^{207}Pb/^{206}Pb$ age equation :

$$\frac{^{207}Pb*}{^{206}Pb*} = \frac{1}{137.88}\frac{e^{\lambda_{235}t} - 1}{e^{\lambda_{238}t} - 1}$$

In other words, we can analyse only Pb isotopes in a particular sample and still be able to obtain an age derived from the U-Pb disintegration series and interpreted in the same way as the U/Pb ages. This specific method requires the subtraction of common lead, however, there is another option to account for it, and that is the $^{207}Pb/^{206}Pb$ isochron.

If we take the two U-Pb isochron equations and divide them by each other, we obtain an equation involving only Pb isotopes:

$$\frac{\left(^{207}Pb/^{204}Pb\right)_{now} - \left(^{207}Pb/^{204}Pb\right)_{original}}{\left(^{206}Pb/^{204}Pb\right)_{now} - \left(^{206}Pb/^{204}Pb\right)_{original}}$$
$$= \frac{1}{137.88} \frac{e^{\lambda_{235}t} - 1}{e^{\lambda_{238}t} - 1}$$

Thus, if we plot our analyses in a diagram of $^{207}Pb/^{204}Pb$ versus $^{206}Pb/^{204}Pb$, we will obtain an isochron whose slope corresponds to the age of the sample.

To conclude, the U-Pb disintegration series are very powerful and versatile, as they can be used in different ways leading to different ages, and can also inform us about the timing of specific events that affected the sample analysed.

The combined U-Pb—K-Ar approach can best be illustrated by the following example. The McLean granitic pluton from the western part of the Grenville orogeny (Ontario, Canada) was first dated by the U-Pb method applied on zircon grains, which yielded an upper intercept age of 1070 ± 7 Ma; the $^{207}Pb/^{206}Pb$ age calculated using the same analyses was 1069 ± 2 Ma [4]; Fig. 4.11). Amphibole, biotite, and potassic feldspar were also extracted and dated by the $^{40}Ar/^{39}Ar$ method (a method derived from the K-Ar method) and the ages obtained were approximately 1060 Ma, 990 Ma, and 790 Ma, respectively (author's data). Closure temperatures were estimated at approximately 500 °C for amphibole, 300 °C for biotite, and 200 °C for

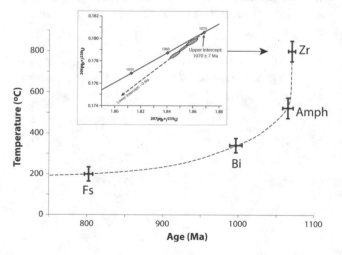

Fig. 4.11 Emplacement age and thermal evolution of the McLean granitic pluton (west Grenville orogeny, Canada), derived from U-Pb dating of zircons (inset) and $^{40}Ar/^{39}Ar$ dating of amphibole, biotite, and potassic feldspar. After a quick initial cooling, the subsequent cooling was much slower, corresponding to the thermal history of the hosting tectonic block (author's own data). This example illustrates vividly that different geochronological methods applied on different minerals will give different ages, which have to be interpreted very carefully

K-Feldspar. Using these data, it was possible to reconstruct the thermal evolution of the McLean pluton (Fig. 4.11): the initial emplacement of the pluton was approximated to the zircon upper intercept age at 1070 Ma. This was followed by rapid cooling, from ca. 800 to ca. 500 °C in about 10 Ma, probably because the host rock into which the pluton was emplaced was considerably cooler. From there on, the cooling of the pluton slowed down significantly: it lost 100 °C from biotite to K-feldspar, a span of approximately 200 Ma. This is a classic example of thermochronology, or the science of studying the thermal evolution of rocks: it has many applications, such orogenic and tectonic studies. Most importantly, it illustrates the principle that any isotopic age corresponds to an event, in most cases the time when the isotopic system was closed, or in other words the time when the rock cooled below the specific closure temperature(s).

4.3 Radiogenic Isotopes and Earth's Evolution

Radiogenic isotopes are a very powerful tool in deciphering Earth's past, not only through geochronology, as seen above, but in many other ways, as well. Principal among those are the study of paleoenvironment, the formation of the atmosphere (as we will see later), the dating of erosion, and the differentiation and evolution of major mantle reservoirs. Here, we will consider the latter, which concerns a very major part of isotope geochemistry efforts.

Let us consider the example of the Rb-Sr parent–daughter pair, where ^{87}Rb disintegrates to ^{87}Sr. Significantly, Rb and Sr have different geochemical behaviours. Rb, is very close in characteristics—and thus in behaviour—to K, which it replaces readily in the crystalline structure of minerals. Indeed, these two elements are always found near each other (e.g., Rb is always enriched in K-rich minerals).

Crucially, Rb is a highly incompatible element, meaning that during any episode of silicate melt coexisting with silicate crystals, Rb will preferentially remain in the silicate melt: it is incompatible with the silicate crystalline structure. Thus, if we have a magma chamber in which some crystals have formed, Rb will avoid those and remain in the silicate melt. If, on the other hand, we are melting some rock due to increased pressure and temperature (or other causes), Rb will readily go into the newly formed melt and leave the crystals quickly. In geology, the processes leading to the formation and separation of various igneous rocks (chief among which are the fractional crystallization and partial melting described above) are collectively called magma differentiation. It is magma differentiation that is the leading cause for the formation of the full range of igneous rocks, from ultramafic, undifferentiated, rocks (e.g., peridotite, picro–basalt, or komatiite), rich in compatible elements, to felsic, or highly differentiated, rocks (e.g., leucogranites, carbonatites, and pegmatites), rich in incompatible elements.

The ^{87}Rb/^{87}Sr geochronology method follows the isochron method, whereby we divide each side of the age equation by the stable and non-radiogenic ^{86}Sr:

$$\left(^{87}Sr/^{86}Sr\right)_{now} = \left(^{87}Sr/^{86}Sr\right)_{original}$$
$$+ \left(^{87}Rb/^{89}Sr\right)_{now}\left(e^{\lambda}\,t - t\right)$$

One corollary of this formulation is that not only do we obtain the age when the rock formed, but also its initial ^{87}Sr/^{86}Sr ratio. That ratio will be directly dependent on the amount of Rb in the rock, as more ^{87}Rb will result in more ^{87}Sr. In other words, higher *initial* ^{87}Sr/^{86}Sr ratios correspond to higher amounts of Rb in the rock (or the reservoir from which it formed), and thus to the degree of magma differentiation (Fig. 4.12). Therefore, we can trace the history of any igneous rock reservoir by using its ^{87}Rb/^{87}Sr age and initial ^{87}Sr/^{86}Sr ratio: we can find out when the rock formed and from which reservoir it formed.

The typical range of initial ^{87}Sr/^{86}Sr ratios on Earth is between 0.699, for very old and completely undifferentiated basalt in the mantle, to approximately 0.82, for young and highly differentiated granites. However, even though most granites will be only up to 0.75, today's highly differentiated Himalayan granites stand at 0.78. Importantly, the main reservoirs on Earth have specific and well known $(^{87}Sr/^{86}Sr)_i$. Meteorites, used as proxy for the initial undifferentiated bulk Earth, have $(^{87}Sr/^{86}Sr)_i$ values of 0.699, while mid-ocean ridge basalts (MORB, roughly corresponding to the upper mantle) have values of 0.702. The

ocean island basalts (OIB, representing the lower mantle), stand at 0.704, and the continents have a $(^{87}Sr/^{86}Sr)_i$ in excess of 0.71 (Fig. 4.13).

Another parent–daughter couple, ^{147}Sm-^{143}Nd, has a very similar behaviour to that of the ^{87}Rb-^{87}Sr pair, but with one crucial exception: Sm is highly compatible, preferentially fractionating into the crystalline phases. Thus, when the ^{143}Nd/^{144}Nd ratio is considered (^{144}Nd being the stable non-radiogenic Nd isotope we normalize by), all observations regarding the ^{87}Sr/^{86}Sr will be reversed. Thus, high ^{143}Nd/^{144}Nd ratios will correspond to lower degrees of differentiation. Ultimately, we can use either of the two systems to identify the degree of differentiation and identify Earth's main reservoirs.

When speaking of main reservoirs, let us consider now those of the mantle. We have access to them via the various volcanic rocks observed in different geodynamic settings. The two major ones are the Mid-Ocean Ridge Basalts (MORB) and the Ocean Island Basalts (OIB, originating from hot spots). However, when we consider the combination of Sr and Pb isotopes (Fig. 4.14), two other reservoirs become apparent, one with high ^{87}Sr/^{86}Sr ratios, called the Enriched Mantle, and another, with high ^{206}Pb/^{204}Pb ratios, called High-μ. Now, μ is equal to the ^{238}U/^{204}Pb ratio and varies from around 7 for the mantle to around 10 for the highly differentiated upper crust; this means that high μ values correspond to high degree of mantle differentiation.

It is somewhat unclear at present what the origins of the Enriched Mantle and High-μ are. Given that high μ values correspond to differentiated rocks from the upper continental crust, one possible interpretation is that the High-μ reservoir corresponds to such rocks that have been incorporated—

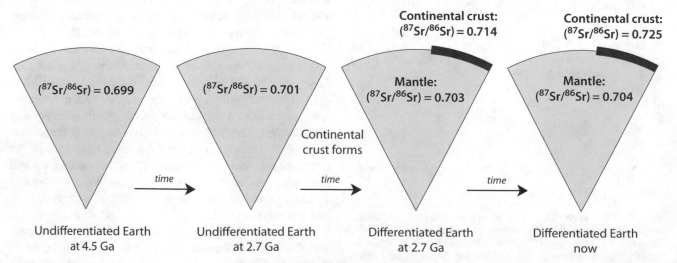

Fig. 4.12 Schematic representation of the evolution of the Earth and the ^{87}Sr/^{86}Sr ratios of the main Earth reservoirs. The ^{87}Sr/^{86}Sr ratios correspond to both differentiation and age

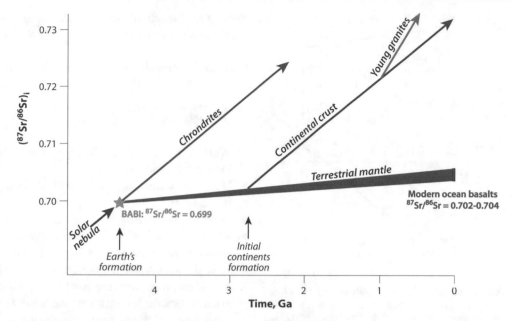

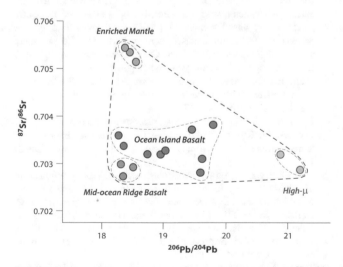

Fig. 4.13 Evolution of the $^{87}Sr/^{86}Sr$ ratios of the main Earth reservoirs in time, starting with the Basaltic Achondrite Best Initial (BABI), with an $^{87}Sr/^{86}Sr$ ratio of 0.699. BABI is considered the best representation of the initial undifferentiated Earth. Chondrites, specific meteorites chemically representing the bulk silicate Earth (BSE), follow their own evolution, whereas Earth's mantle follows a different evolution. The continents represent a second major reservoir, enriched in Rb as more differentiated and having their own Sr isotopic evolution. Younger granites, strongly differentiated, will have the highest $^{87}Sr/^{86}Sr$ ratio, going up to 0.82

through subduction and continental collision—back into the mantle. It is possible that a similar explanation can also be applied to the Enriched Mantle.

Let us now complicate things a bit further and include Nd isotopes in our considerations about the main mantle reservoirs. As mentioned, Nd isotopes should behave in a similar but opposed way to the Sr and Pb isotopes: we will notice that while high $^{87}Sr/^{86}Sr$ and $^{206}Pb/^{204}Pb$ ratios indicate high degree of differentiation, a high $^{143}Nd/^{144}Nd$ ratios correspond to a low degree of differentiation. When we plot all three ratios in the same three-dimensional diagram (Fig. 4.15), all observations fall on or very near to the same plane, with the three end-members of mantle reservoirs still clearly visible.

In conclusion, the radiogenic isotopes of several elements —chief among which are Sr, Pb, and Nd—are not only a very powerful tool in distinguishing the different reservoirs of the Earth's mantle, but also a very useful tool in studying the evolution of any rock or rock type. Thus, any given rock will have a very distinctive and specific radiogenic isotopic ratio. This ratio will correspond to its age, but also to its formation and evolution, for instance its degree of differentiation. By combining such information as bulk rock chemical composition, stable isotopes and radiogenic isotopic compositions and, of course, geological observations, we can obtain very clear and specific ideas about the formation and evolution of any given rock.

Fig. 4.14 The main mantle reservoirs, represented in a $^{87}Sr/^{86}Sr$ versus $^{206}Pb/^{204}Pb$ diagram. Three distinct reservoirs become apparent: the Enriched Mantle (high $^{87}Sr/^{86}Sr$ ratios), the MORB (low $^{87}Sr/^{86}Sr$ and $^{206}Pb/^{204}Pb$), and the High-μ, with high $^{206}Pb/^{204}Pb$ ratios. The Ocean Island Basalts—normally originations from hot spots—seem to be a mixture of the three sources. Data from White [8], Zindler and Hart [10], and Hauri et al. [5]

Further Reading

The study of rocks and the Earth has been strongly relying on isotopes for many decades, and therefore there is no shortage of information available. A few select books include:

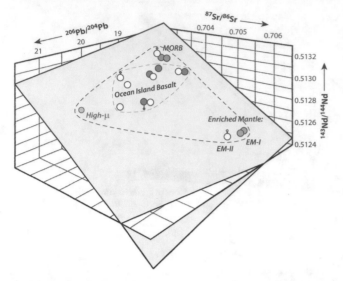

Fig. 4.15 The three mantle reservoir end-members and the Ocean Island Basalts, which represent a mixture of the end-members, are plotted in a three-dimensional diagram of $^{87}Sr/^{86}Sr$, $^{206}Pb/^{204}Pb$, and $^{143}Nd/^{144}Nd$ ratios. All observations fall on or near the same plane (clear circles are below the plane, solid circles are on it or above it). After Zindler et al. [9], modified

Isotopes in the Earth Sciences, H.-G. Attendorn and R. Bowen, Springer Science, 1988, ISBN 978-041-253710-3

Stable Isotope Geochemistry: a Tribute to Samuel Epstein, H.P. Taylor, J.R. O'Neil, and I.R. Kaplan, Editors. The Geochemical Society Special Publication #3, ISBN 0-941809-02-1. Of particular interest here are Parts E and F.

Radiogenic Isotope Geology, A.P. Dickin, Cambridge University Press, 2005, ISBN 0-521-82316-1. A valuable book that will be immensely helpful to those studying radiogenic isotopes in Earth sciences.

Isotope Geology, C. Allegre, Cambridge University Press, 2008, ISBN 978-0-521-86228-8.

Isotope Geochemistry, W.M. White, Willey, 2015, ISBN 978-0-470-65670-9.

Stable Isotope Geochemistry, J. Hoefs, Springer-Verlag, 1997, ISBN 3-540-61126-6.

Geochronology and Thermochronology by the $^{40}Ar/^{39}Ar$ Method, I. McDougall and T.M. Harrison, Oxford University Press, 1999, ISBN 0-19-510920-1. This is the bible of $^{40}Ar/^{39}Ar$ geochronology. If you need one book about this method, this is the one to choose.

Noble Gas Geochemistry, M. Ozima and F.A. Podosek, Cambridge University Press, 2002, ISBN 0-521-80366-7.

Isotopes, Principles and Applications, G. Faure and T.M. Mensing, Willey, 2005, ISBN 978-0-471-38437-3.

Questions

- What is the most common type of fractionation in geological processes and why?
- How are water isotopes modified during water–rock interaction?
- In what way can isotopes be used in mineral exploration?
- What is the basic equation used in geochronology? What are the main isotopes used in geochronology?
- How do we correct for common lead? What is an isochrone and how does the isochrone method work?
- What is closure temperature and what factors affect it? What practical application do we use it for?
- What are Concordia and Discordia and how are they used?
- How can radiogenic isotopes be used to study Earth's evolution and the main mantle reservoirs?

References

1. Bottinga Y, Javoy M (1973) Comments on oxygen isotope geothermometry. Earth Planet Sci Lett 20:250–265
2. Brownlow AH (1979) Geochemistry. Prentice-Hall, 498 p
3. Clayton RN, O'Neil JR, Mayeda TK (1972) Oxygen isotopic exchange between quartz and water. J Geophys Res 77:3057–3067
4. Davidson A, van Breemen O (2000) Late grenvillian granite plutons in the central metasedimentary belt, grenville province, southeastern ontario. Geol Surv Can, current research 2000-F5, radiogenic age and isotopic studies: Report 13, 9 p
5. Hauri EH, Shimizu N, Dieu JJ, Hart SR (1993) Evidence for hotspot-related carbonatite metasomatism in the oceanic upper mantle. Nature 365:221–227
6. Kyser TK (1987) Equilibrium fractionation factors for stable isotopes. In: Kyser TK (ed) Stable isotope geochemistry of low temperature processes. Mineral. Assoc Can Shourrt Course Handb 13
7. Polito P, Kyser K, Lawie D, Cook S, Oates C (2007) Application of sulphur isotopes to discriminate Cu–Zn VHMS mineralization from barren Fe sulphide mineralization in the greenschist to granulite facies Flin Flon-Snow Lake–Hargrave River region, Manitoba, Canada. Geochem: Explor Environ, Anal 7:129–138
8. White WM (1985) Sources of oceanic basalts: radiogenic isotope evidence. Geology 13:115–118
9. Zindler A, Jagoutz E, Goldstein S (1982) Nd, Sr, and Pb isotopic systematics in a three-component mantle: a new perspective. Nature 298:519–523
10. Zindler A, Hart SR (1986) Chemical geodynamics. Annu Rev Earth Planet Sci 14:493–571

5.1 The Complexity of the Biosphere

Studying the living organisms, both plants and animals, their past and present mode of life and interactions, the way they function, can be a very complex task that broadly encompasses the sciences of biology, zoology, and ecology. This task is made very difficult by the constant and fast-paced changes and interactions between the organisms we study and their environment. The organisms also exist within a very complex framework, or web, of relationships involving energy, food, and the natural environment.

The last two or three decades have seen that isotopic applications to the living world emerge as the fastest growing application of isotopes (after the 50–80 s during which it was all about geological applications). The reasons for this are multiple, including but not limited to technological and analytical developments and better understanding of isotopic behaviour in the living organisms domain. Indeed, all processes involving living organisms induce strong variation in stable isotopes, as the fractionations observed are almost always kinetic fractionations. As we noted earlier, in Chap. 1, kinetic fractionations are more difficult to reproduce, understand, and quantify than equilibrium fractionations. Additionally, a great many factors affect the isotopic compositions of living organisms, such as the health and behaviour of the organism, environmental variations, including availability of food, position within the food web, and diet modifications. The good news is that the effects of these factors are better and better known, which allows us to gather—mostly through the use of light stable isotopes—a great deal of information about a living organism.

5.2 Plants Respiration and Carbon Fixation

Let us consider the relatively simple example of plant respiration. In the most general form, plants use the sun's energy (and CO_2 and H_2O, from the air and ground), to generate oxygen and glucose, in the process that we call photosynthesis:

$$6CO_2 + 12H_2O + \text{sun's energy} \rightarrow C_6H_{12}O_6 + 6O_2 + 6H_2O$$

This process occurs in plants' leaves, through specific organs called *stoma*, consisting of an opening, or pore, and a pair of guard cells (Fig. 5.1). The purpose of the stoma is to allow CO_2 to enter and for the O_2 to exit the sub-stomatal cavity. The mesophyll (or chloroplast) then acts to assimilate CO_2 and fix carbon into the plant's cells.

The processes of carbon assimilation can be fairly complex. Most commonly, approximately 90% of plants (such as most broadleaf plants and plants in temperate zones) use the so-called Calvin cycle, where CO_2 is fixed through the use of the RuBisCO (*ribulose-1,5-bisphosphate carboxylase/oxygenase*) in the mesophyll. The RuBisCO is an enzyme that works in a cycle of three phases: firstly, it fixes carbon in phosphoglycerate ($C_3H_7O_7P$), which is then reduced, and finally ribulose is regenerated to serve again. This is not the most efficient method of carbon fixation, as the RuBisCO does not have a high affinity for CO_2, but also because some oxygen is fixed, which results in some amount of energy being wasted to release it. As a result, the C_3 plants (called so because of the three carbon atoms fixed in the first stable intermediate molecule of the cycle, phosphoglycerate) require higher ambient CO_2 concentrations and higher stomatal apertures.

An improvement on this carbon fixation method is seen in so-called C_4 plants (approximately 10% of all plants). In C_4 plants, the first stable intermediate molecule to fix CO_2 contains four carbon atoms is oxaloacetate ($C_4H_4O_5$), which then releases CO_2 to the Calvin cycle [4]. This is a much more efficient process and appeared relatively late in the evolution of plant life on Earth, sometime in the late Miocene, or approximately 6–8 million years ago. Typical examples are tropical grasses, such as sugar cane and maize, but there are also broadleaf plants that use C_4 carbon fixation.

© Springer Nature Switzerland AG 2020
P. Alexandre, *Isotopes and the Natural Environment*, Springer Textbooks in Earth Sciences, Geography and Environment, https://doi.org/10.1007/978-3-030-33652-3_5

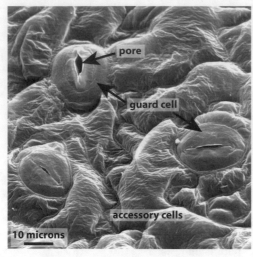

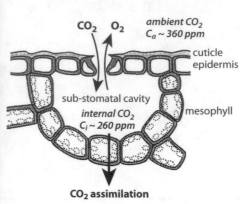

Fig. 5.1 Stoma, the plant organ responsible for plant respiration and consisting of a pore and two guard cells. The guard cells are responsible for the opening (during the day) and closing (during the night) of the pore and for regulating the amount of gases that will go through. The

amounts (or partial pressure) of CO_2 outside (ambient) or inside (internal) will depend on many environmental factors and on the extent of the opening of the pore

(A third method of carbon fixation exists, but is exceedingly rare and seen only in some cacti and in pineapple. It is called the CAM, or *Crassulacean Acid Metabolism*. In this process, CO_2 entering the stomata during the night is converted into organic acids, which release CO_2 during the day, when the stoma is closed. These are the only plants that "breathe" during the night. We needn't worry too much about them here.)

As far as isotopes are concerned, there is a strong kinetic fractionation that occurs during plant respiration and carbon fixation. In each case and during any process, $^{12}CO_2$ is preferred by plants over $^{13}CO_2$ during CO_2 intake, assimilation by RuBisCO, and conversion of CO_2 into organic acids. Further fractionation will occur during the formation of more complex compounds, such as cellulose or lipids. The reason for this preference is that $^{12}CO_2$ is a smaller and lighter molecule than $^{13}CO_2$ and, most importantly, carbon is slightly less strongly bound in the $^{12}CO_2$ molecule than in the $^{13}CO_2$ one. This difference makes it more energy efficient for the plant to assimilate ^{12}C than ^{13}C.

These kinetic fractionations are fairly large: the typical *discrimination* for C_3 plants (or the difference between the $\delta^{13}C$ of the air and that of the plant) is approximately 18‰. The air $\delta^{13}C$ is typically -8‰ and that of C3 plants is, on average, −26‰ (even though it will vary from approximately −22 to −32‰; Fig. 5.2). Significantly, C_4 plants are more efficient in CO_2 intake and carbon fixation, resulting in stronger discrimination against ^{13}C: they display lower fractionation relative to the air, about 5‰, and their carbon isotopic composition is approximately −13‰ (−10 to 15‰; Fig. 5.2).

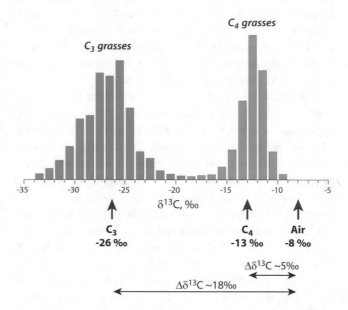

Fig. 5.2 Carbon isotope discrimination between C_3 and C_4 plants: C4 plants are more efficient in carbon intake and fixation and discriminate more against ^{13}C (Ehleringer and Cerling [3])

The appearance of C_4 plants has been documented in the fossil record. Carbonate minerals were extracted from paleo-soils in Pakistan [1] and a clear jump from about −10‰ to about 2‰ was observed in the late Miocene, at about 7 million years ago (Fig. 5.3): this was interpreted as grasses in the paleo-soil switching from mostly C_3 to predominantly C_4 method of CO_2 intake and carbon fixation. The difference observed between soils, approximately 12‰, is very similar to that between C_3 and C_4 plants.

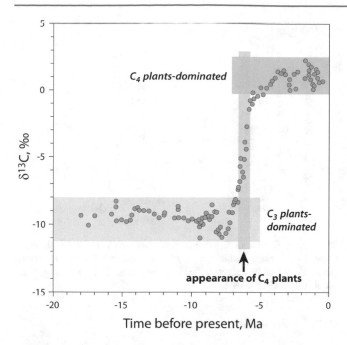

Fig. 5.3 Change of fossil soil carbonates' carbon isotopic composition, occurring at approximately 8 to 5 million years ago and interpreted as switching from C_3- to C_4-dominated plants (after Cerling et al. [1], modified)

We notice that there is a significant variation in carbon isotopic composition in both C_3 and C_4 plants (Fig. 5.2). This is because several other factors will affect the plant's isotopic composition, main among which are water availability and water use efficiency, photosynthetic capacity, RuBisCO activity, amount of sunshine, and stomatal conductance. The RuBisCO activity, A, is directly related to stomatal conductance (g) and the ratio of CO_2 concentrations in the air and in the sub-stomatal cavity (C_a/C_i):

$$A = g(C_a/C_i)$$

In other words, when stomatal conductance is low compared with RuBisCO activity, C_a/C_i is high; when stomatal conductance is high compared with RuBisCO activity, C_a/C_i is low. When C_a/C_i changes, so does the isotopic discrimination between CO_2 in the air and that of the plant ($\Delta = \delta^{13}C_{air} - \delta^{13}C_{plant}$) to a significant degree (Fig. 5.4a).

In general terms, the RuBisCO activity, A, reflects variations in plant health or nutrition availability (such as fertilization); the stomatal conductance, g, reflects the availability of water (Fig. 5.4b); and together the ratio A/g reflects the overall water use efficiency. Significantly, these are clearly reflected in the plant's carbon (but also nitrogen) isotopic composition and can be measured.

Another environmental factor is the amount of sunshine a plant receives: it is directly related to increased RuBisCO activity, leading to higher Ca/Ci ratios and to lower

$\delta^{13}C_{air} - \delta^{13}C_{plant}$ discrimination. Finally, the overall photosynthetic capacity of a plant is also directly correlated to higher $\delta^{13}C_{plant}$ values and thus to lower $\delta^{13}C_{air} - \delta^{13}C_{plant}$ discrimination.

In other words, at any point, we can have a clear idea about a plant's health and its environmental conditions by analyzing its carbon isotopic composition: this is a very powerful tool available for plant scientists and ecologists.

> **Fun fact: when carbon isotopes help food industry**
> Canada is many things, but the maple tree is of particular importance to Canadians: it is on the country's flag, after all! And let us also not forget about maple syrup: Canada is by far the largest producer of this delicious syrup (resulting from the evaporation of maple tree sap), which is why it is treated with supreme seriousness there by both producers and the government. Now, maple syrup is produced manually and is energy intensive, resulting in a relatively high cost. Some unscrupulous people have been tempted to dilute maple syrup with the much cheaper corn syrup to reap a profit. However, they did not study isotopes in plants! Maple is a C_3 plant, which means that the maple syrup will have a $\delta^{13}C$ of about −26‰ (carbon isotopes will not be affected during maple syrup production). However, corn is a C_4 plant, resulting in corn syrup having a $\delta^{13}C$ of about −10‰. Thus, even small amounts of corn syrup in maple syrup, less than 10%, can be easily detected using carbon isotopes (Fig. 5.5).
>
> Carbon isotopes, and also the isotopes of other elements, are commonly used to detect fraud in the food industry (food forensics), as they will faithfully reflect the isotopic compositions of the individual produce used, conditioned mostly by environmental factors. A good example is wine as its Sr, H, and O isotopic compositions will faithfully and precisely reflect the place where the grape was grown, preventing unscrupulous producers from claiming that their wine comes from another, more prestigious—and thus more expensive—locality. We will see much more about forensics, including wine provenance, in Chap. 9.

5.3 Animal Ecology

Animal ecology is a very vast and complicated field, and we can only scratch the surface of the wealth of knowledge available. The science of ecology has many branches, the main of which are animal ecology and migration (nutrition,

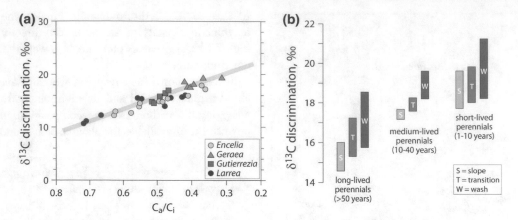

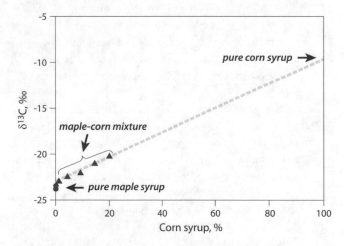

Fig. 5.4 Carbon isotopes' discrimination between air and plants ($\Delta = \delta^{13}C_{air}—\delta^{13}C_{plant}$, ‰), as a function of plant stomatal conductance (**a**), reflected in the CO_2 ambient and internal concentrations (C_a/C_i) for four species of flowers in the southern USA, and the water

availability (**b**) for different longevity categories perennial plants. In both cases, plant health and environmental conditions are clearly reflected in plants' carbon isotopic composition. Data from Ehleringer et al. [5] and Farquhar et al. [7]

5.3.1 Food Sources

In both animals and people (as we will see later, in Chap. 7), isotopically, "You are what you eat": diet is clearly reflected by the carbon and nitrogen isotopes in different tissues. In other words, the isotopic signature of a consumer's tissues reflects isotopic signatures of food sources, proportional to their dietary contribution. Let us take bison, for instance, and its diet. Imagine a bison that roams the Great Plains of the USA Midwest and eats as it pleases, its diet consisting of C_3 and C_4 grasses that we discussed earlier in this chapter. The question then is, what are the proportions of C_3 and C_4 grasses in bison's diet? Our known factors are the bison's carbon isotopic composition, measured in hair (e.g., $\delta^{13}C = -21‰$), and the carbon isotopic compositions of C_3 and C_4 grasses (-26 and $-13‰$, respectively). We already saw how the quantification of isotopic mixture works (Chap. 3), and so, applied to this case, we can rewrite the isotopic mixture equation like this

$$\delta^{13}C_{BISON} = \delta^{13}C_{C3}f_{C3} + \delta^{13}C_{C4}f_{C4}$$
$$= \delta^{13}C_{C3}f_{C3} + \delta^{13}C_{C4}(1 - f_{C3})$$

Here, the only unknown will be the proportion of C_3 plants (f_{C3}). After applying the numbers, we come to $f_{C3} = 0.615$, meaning that 61.5% of the bison's diet is C_3 grasses and 38.5% is C_4 grasses.

This is all very neat and tidy, but somewhat simplistic. (However, it works very well when calculating *proportions* of food sources.) Firstly, we have to speak of *assimilated* diet, as that is what is going to be reflected isotopically in the consumer's tissues, and not of *ingested* diet. Secondly, the isotopic compositions must be adjusted for tissue-diet

Fig. 5.5 Carbon isotopic composition of pure maple syrup (sourced from maple, a C_3 plant) and of maple syrup–corn syrup mixture (corn is a C_4 plant). Even small amounts of corn syrup added into maple syrup, even under 10%, are clearly reflected in the carbon isotopic composition of the mixture

trophic relationships, foraging strategies, migration), aquatic ecosystems (estuarine ecology, limnology, benthic ecology), soil and microbial processes (biogeochemical cycling, and soil formation and evolution, decomposition, microbial biochemistry). Also included are paleoecology (climate change and environmental reconstruction, which we will consider in the next chapter, and archaeology, subject of the Chap. 8), and large-scale ecosystem processes (gas fluxes and atmospheric–terrestrial–marine connections). Importantly, isotopes are used in all of these areas, and we will consider a few examples here that specifically relate to food sources, nutrition evolution, and migration patterns.

discrimination and, at any point, we must ask ourselves how isotopes will be affected when transiting through the different tissues of the consumer's body. There are two main and opposite effects that are at work during digestion and assimilation of nutrition. Firstly, heavy isotopes will be discriminated against during digestion, as they form stronger bonds and the molecules containing more heavy isotopes are thus more difficult to break down and digest (and become assimilated diet). This is particularly visible with microorganisms and plants, where nutrition molecules must be completely broken down for individual elements to be released and be a part of the diet. Secondly, light isotopes are discriminated against during waste elimination. Heavy isotopes, forming stronger bonds, will be preferentially retained in the consumer's tissues, while light isotopes will be preferentially eliminated. These are not very strong effects, but are quantifiable, and always result in the consumer having a slightly heavier isotopic composition (i.e., slightly enriched in the heavy isotope) that is its food source. In other words, we can change the initial statement to say that, isotopically, "You are what you eat, *plus a few per mil.*" (‰ ≡ permil) We call this overall effect the *trophic shift*.

In most general terms, the word trophic means "relating to feeding and nutrition", but we will take it to mean "relating to feeding and nutrition, with regards to a relative position in the food web". (Let us not speak of food *chain*, which is a bit too simplistic: in reality, most everything eats something else and is food to something else, which creates a very complex relationship between species.) In our case, the trophic shift will mean the slight increase in delta values as species are positioned at a higher level in the food web (Fig. 5.6), underlining the earlier statement ("You are what you eat, plus a few permil").

5.3.2 Paleoecology

Genyornis newtoni, an extinct Australian flightless bird, is a very interesting and revealing proxy for the effects of humans on their environment and, more specifically, the fauna. It had co-existed with other flightless birds, the emu (*Dromaius novaehollandiae*) and the cassowary (*Casuarius casuarius*), for more than 100,000 years, but disappeared about 45–50 thousand years ago, along with most large terrestrial genera (85% of Australian terrestrial species with a body weight above 44 kg disappeared in the Late Pleistocene). Using stable isotopes—carbon ones in particular—has been very useful in deciphering the story of this mass extinction and has provided some hints as to its cause [8].

In this case, the main material used was the *Genyornis* and emu fossil eggshells. Given that they contain a significant amount of carbon, they can provide both ^{14}C ages and $\delta^{13}C$ values. The carbon isotopic signature of the eggshells is conditioned by the preferential diet, which in turn is

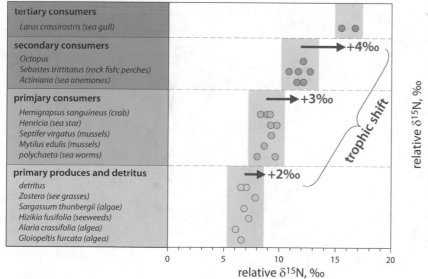 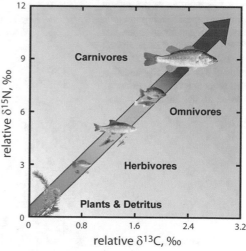

Fig. 5.6 Left: Trophic shift seen in the change of nitrogen isotopic composition of marine species, from primary producers and detritus to tertiary consumers (pure carnivores). Right: illustration of trophic shift of marine species for both carbon and nitrogen isotopes. The isotopic effects of trophic shift are small but measurable

affected by climate and other environmental factors. There is a further complication, alluded to earlier, in that there will be a significant fractionation between the assimilated diet and the eggshell ($\Delta = \delta^{13}C_{EGGSHELL} - \delta^{13}C_{DIET}$), which has been quantified at between 10 ± 2 and 14 ± 2‰ for different modern and fossil ratites. The main climate factor affecting these species is precipitation. It has been observed that, as precipitation in summer months increases, modern emu have an increased diet of C_4 grasses, which in turn increases the $\delta^{13}C$ values of their eggs; inversely, as precipitation decreases, they consume increased amounts of C_3 flowers, seeds, and shoots, which is reflected in decreased $\delta^{13}C_{EGGSHELL}$. It is expected that similar variations affected the extinct *Genyornis*. We also noted earlier the influence of precipitation on plants' carbon isotopic composition: in Australia, $\delta^{13}C$ of C_3 plants will vary between approximately -26‰ in the dry months to approximately -30‰ in the high precipitation periods.

When we observe the evolution of carbon isotopic compositions for emu and *Genyornis* eggshells, we see a marked change in the emu diet just at the time when *Genyornis* became extinct. Before approximately 45 thousand years ago, both species consumed a C_4-dominated diet, and were slowly evolving towards an increased proportion of C_3 plants (Fig. 5.7). The emu appears to have been more of a generalist, with a more varied diet ($\delta^{13}C_{EGGSHELL}$ between -15 and -2‰), compared to the more restricted range of

carbon isotopic compositions of the specialist *Genyornis* (-11 to -1‰). Importantly, both species were in a period of switching their diet towards more C_3 plants.

Three sets of factors may have led to the demise of *Genyornis*: competition, predation, or climate change. The last factor, climate change, seems to not have been as significant, because climate has not been documented to change substantially at the time of *Genyornis* extinction. Was it predation or competition, then? We are not certain, but the arrival of humans in Australia at the same time must have been a factor. People were burning bush for clearing, thus restricting the food sources of many species. Some of these, such as emu, were more opportunistic feeders and thus were able to survive by adapting to the available (predominantly C_3 plants) food sources. Species with more restricted dietary requirements, such as *Genyornis*, were at greatest risk and eventually became extinct.

A somewhat similar example is that of the Adélie penguin (*Pygoscelis adeliae*) and the evolution of their diet [6]. Eggshells from nearly 40 thousand years before present were collected from several Antarctica locations, dated by the ^{14}C method, and analyzed for carbon, nitrogen, and oxygen isotopes. The Adélie penguin has a varied diet, consisting of krill, three species of fish, and an unknown source characterized by high $\delta^{13}C$ and low $\delta^{15}N$ (Fig. 5.8). Importantly, the carbon, nitrogen, and oxygen isotopic compositions of Adélie penguin eggshells changed sharply somewhere

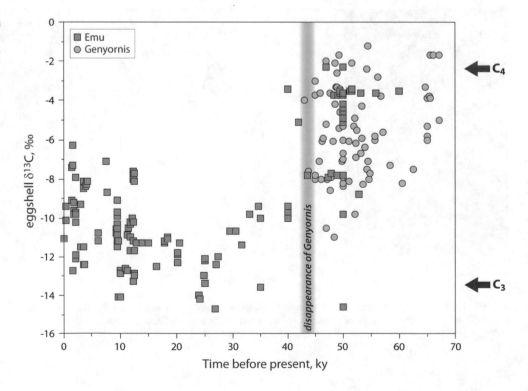

Fig. 5.7 Evolution of the carbon isotopic composition of emu and *Genyornis* eggshells over the last 70 thousand years. The position of eggshell compositions produced by mainly C_3- and C_4-based diets is indicated. From Miller et al. [8], modified

Fig. 5.8 Carbon and nitrogen isotopic composition of Adélie penguin eggshells from different Antarctica locations. Their main diet consists of krill, three species of fish, and an unknown source. Data from Emslie and Patterson [6]

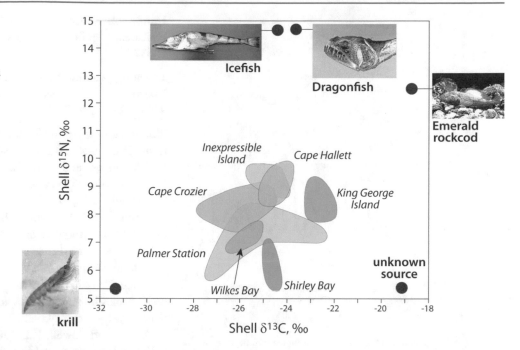

between 150 and 200 years ago: $\delta^{13}C$ decreased from around −10 to −20‰, $\delta^{15}N$ decreased from −12 to −6‰, and $\delta^{18}O$ increased from −9 to −6‰.

The changes of Adélie penguin eggshells' isotopic compositions most likely reflect changes in diet and specifically reflect an increase in the diet of the proportion of krill, which has a low $\delta^{13}C$ and $\delta^{15}N$ (Fig. 5.8). (As seen earlier, the three fish species have high $\delta^{13}C$ and $\delta^{15}N$ values, reflecting their high trophic level.) The increased $\delta^{18}O$ is also consistent with increased krill consumption, as that would mean an increased intake of seawater. This interpretation is fairly straightforward, but why the increased consumption of krill? Well, likely because krill became much more readily available; but there again, we have to ask, why was krill made more available? Here, a brief history of seal and whale hunting in the region might provide some useful clues. Sealing began on the Falklands and on South Georgia Island in the late eighteenth century, in 1766 and 1786, respectively. Within half a century, by the mid-1820s, an estimated 3.5 million skins were taken (yes, millions). The result was the near extinction of seals in the South Atlantic Ocean, but also a very significant depletion of whales in this part of the world, by an estimated 90–95%. Thus, krill, which was the main dietary source of both seals and whales, became readily available and provoked an ecological response in the Adélie penguin: the shift in diet from fish to krill, reflected in the isotopic composition of their eggshells.

5.3.3 Migration Patterns

Between 2000 and 2002, an international team tracked the movements of a small group of elephants (*Loxodonta africana*) in the Samburu National Reserve, Northern Kenya, in the attempt to study their migration patterns and diet changes [2]. The elephants' movements were tracked by GPS radio collars, sending one signal per hour, and visual observation. There were seven elephants in the study, four female and three male.

Six of the seven elephants were well behaved: they were observed in Samburu National Reserve or in the near vicinity, which was corroborated by the GPS information. However, one elephant, an approximately 40-year-old bull elephant (prosaically named B1013), misbehaved: the GPS tracking revealed that he made three major trips, shifting from the arid lowlands of Samburu National Reserve to the mesic Imenti upland forest near Mt. Kenya and back. Each range shift, corresponding to a straight-line distance of more than 40 km, was accomplished in less than 15 h. B1013 did this for the availability of food: the periods he spent in the lowlands correspond to the mid- and late rainy season, and periods spent in the forest were during the dry season, when there was more food in the more humid Imenti upland forest. This is fairly logical, but the question is, can we observe these range changes in the stable isotopic composition of the elephant?

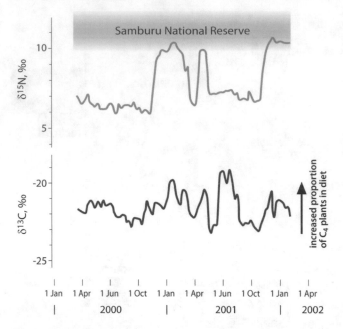

Fig. 5.9 Carbon and nitrogen isotopic composition of hair of the B1013 male elephant, clearly representing the shifts of range and diet. The estimated proportion of C_4 plants varies from 0 to 50%, but does not coincide with migration. Data from Cerling et al. [2]

Hair samples were taken at regular intervals from B1013 and analyzed for carbon and nitrogen isotopes. The results (Fig. 5.9) clearly show not only the shift in range, but also in diet. B1013's diet consisted mainly of C_3 plants, but it seems to have included as much as 50% C_4 plants, which did not necessarily with his stays at the Imenti upland forest.

Further Reading

Fortunately, there are some excellent books and journal articles on the topic of isotopes in living beings. Here is a brief selection of these:

Stable Isotope Ecology, B. Fry, Springer, 2008, ISBN 978-0-387-30513-4.

Stable Isotopes in Ecology and Environmental Science, R. Michener and K. Lajtha, Editors, Blackwell, 2007, ISBN 978-1-4051-2680-9. This book is an absolute gem: highly recommended.

Stable isotopes in Ecological Research, P.W. Rundel, J.R. Ehleringer, and K.A. Nagy, Editors, Springer, 1988, ISBN 978-1-4612-8127-6.

Stable Isotopes as Indicators of Ecological Change, T.E. Dawson, and R. Siegwlof, Editors, Academic Press, 2011, ISBN 978-0-080-55100-5.

Handbook of Environmental Isotope Geochemistry, M Baskaran, Springer Science, 2011, ISBN 978-3-642-10637-8.

Tracking Animal Migration with Stable Isotopes, K.A. Hobson and L.I. Wassenaar, Academic Press, 2018, ISBN 978-0-128-14724-5.

Stable Isotopes and Biosphere - Atmosphere Interactions: Processes and Biological Controls, L.B. Flanagan, J. R. Ehleringer, and D.E. Pataki, Editors, Elsevier, 2004, ISBN 978-008-052528-0.

O'Connell. T.C., Hedges, R.E.M., Healey, M.A., and Simpson, A.H.R.W. (2001) Isotopic composition of hair, nail and bone: modern analyses. Journal of Archeological Science, 28, 1247–1255.

Bol, R., and Pflieger, C. (2002) Stable isotope analyses (^{13}C, ^{15}N, and ^{34}S) analysis of the hair of modern humans and their domestic animals. Rapid Communications in Mass Spectrometry, 16, 2195–2200.

Still, C.J., and Powell, R.L. (2010) Continental-scale distributions of vegetation stable carbon isotope ratios. In B. West et al. (eds.), Isoscapes: Understanding Movement, Pattern, and Process on Earth Through Isotope Mapping. Springer Science + Business Media B.V. 2010, pp. 179–193.

Questions

- What type of fractionation is most common in the biosphere?
- What are the two main mechanisms of carbon fixation and how do they affect carbon isotopes fractionation? When did the second type appear and how do we know that?
- How can we derive the diet or food sources of an animal (or a human)? The isotopes of which elements would we use?
- What is trophic shift? How does it work and on what basic principle is it based?
- How can we study the diet of an extinct animal? What factors must we take into account?

References

1. Cerling TE, Wang Y, Quade J (1993) Expansion of C_4 ecosystems as an indicator of global ecological change in the late miocene. Nature 361:344–345
2. Cerling TE, Wittemyer G, Rasmussen HB, Vollrath F, Cerling CE, Robinson TJ, Douglas-Hamilton I (2006) Stable isotopes in elephant hair document migration patterns and diet changes. Proc Natl Acad Sci USA 103:371–373

3. Ehleringer JR, Cerling TE (2002) C_3 and C_4 photosynthesis. In: Mooney HA, Canadell JG (eds) Encyclopedia of Global Environmental Change, Editor-in-chief T. Munn, vol 2, The Earth system: biological and ecological dimensions of global environmental. Wiley, Chichester, pp 186–190

4. Ehleringer JR, Cerling TE, Helliker BR (1997) C_4 photosynthesis, at-mospheric CO2, and climate. Oecologia 112:285–299

5. Ehleringer JR, Hall AE, Farquhar GD (1993) Stable isotopes and plant carbon/water relations. Academic Press, San Diego, California, USA

6. Emslie SD, Patterson WP (2007) Abrupt recent shift in $\delta^{13}C$ and $\delta^{15}N$ values in Adélie penguin eggshell in Antarctica. Proc Natl Acad Sci USA 104:11666–11669

7. Farquhar GD, Ehleringer JR, Hubick KT (1989) Carbon isotope discrimination and photosynthesis. Annu Rev Plant Physiol Plant Mol Biol 40:503–537

8. Miller GH, Magee JW, Johnson BJ, Fogel ML, Spooner NA, McCulloch MT, Ayliffe LK (1999) Pleistocene extinction of *Genyornis newtoni*: human impact on Australian megafauna. Science 283:205–208

Atmosphere and Climate

The atmosphere is a complex and dynamic component of our natural environment; it has intricate and multiple interactions with the other "spheres" of the Earth, including the geosphere, the hydrosphere, and the biosphere. Significantly, all interactions and processes involving the atmosphere will be reflected in the isotopic composition of the elements composing it, not the least of which is the modification of its physical characteristics, specifically temperature, or more prosaically, climate change. Here, we will consider isotopes in the atmosphere and how isotopes reflect climate change.

6.1 Isotopes and the Atmosphere

Let us examine first the composition of the atmosphere. Simplistically, it is made of 78% N_2, 21% O_2, and 1% Ar, plus trace amounts of several other elements (Table 6.1). This composition is the result of a long evolution and has changed multiple times: indeed, we are at our fourth atmosphere.

A (Very Brief) History of Earth's Atmosphere *Atmosphere 1* was composed almost uniquely by He and H, inherited from the solar nebula during the formation of the Earth. These were blown away, by solar winds, within the first 100–200 Ma of Earth's history, during the *T-Tauri* stage and ended up in the outer reaches of the solar system, where they coalesced to form the giant gas planets Jupiter and Saturn (made of 74% H, 25% He, and trace amounts of H_2O, CH_4, and NH_3).

Atmosphere 2 existed between approximately 4.4 and 2.7 Ga and corresponded to a strong volcanic activity that was related to the formation of Earth's primitive crust. It was predominantly made of CO_2 and H_2O, with no oxygen present. As the air cooled, CO_2 became more dissolved in the ocean and precipitated in the form of carbonates. It was also rather warm, with the average surface temperature of about 70 °C.

Atmosphere 3 was a transitional one: the first significant life-forms, cyanobacteria, started to consume CO_2 and release O_2. This caused the transition from an anoxic to oxic state of the atmosphere. As plants appeared and started to develop, more CO_2 was sequestered mostly as coal, limestone, and shells. As plants developed further and increased in mass, they sequestered greater and greater amounts of CO_2 and produced more and more O_2. At around 1.8 Ga, with an increased amount of plant mass, nitrogen began to be released into the atmosphere by aerobic nitrification.

Atmosphere 4 is our atmosphere of today, and is generally N and O dominated. Atmosphere 4 is also the direct result of the existence of life on Earth: on Venus and Mars, where no known organic life-forms exist, the atmosphere is composed mostly of CO_2 (96.5% and 95.3%, respectively).

We can also talk about the vertical structure of the atmosphere and its different layers (the troposphere, stratosphere, mesosphere, and thermosphere) and their characteristics, specifically temperature, air dynamics, and composition. From our point of view, we will only focus on the fact that 99.9% of the mass of the atmosphere is found in the two lower layers, troposphere and stratosphere, with the former being rather turbulent and the latter rather calm. Another important feature of the atmosphere is the ozone layer, which is mainly situated in the lower stratosphere, at approximately 20–30 km, and contains approximately 30 times more ozone (O_3) than elsewhere in the atmosphere. The ozone layer is paramount to the survival of life on Earth, as it absorbs most of the harmful ultraviolet radiation that comes from the Sun.

Since oxygen is the second most abundant element in the atmosphere, scientists wanted to study its isotopic composition very early on. They used spaceships, airplanes, and helium balloons, to collect O_2 and CO_2 samples from

© Springer Nature Switzerland AG 2020
P. Alexandre, *Isotopes and the Natural Environment*, Springer Textbooks in Earth Sciences,
Geography and Environment, https://doi.org/10.1007/978-3-030-33652-3_6

Table 6.1 Current composition of the atmosphere and the isotopic composition of some of its main constituents

Gas	Volume %	Current isotopic composition
Main gases		
N_2	78.084	$\delta^{15}N = 0‰$ (by definition)
O_2	20.948	$\delta^{18}O = 23.5‰$
Ar	0.934	$^{40}Ar/^{36}Ar = 295.5$
Trace gases		
CO_2^a	0.036	$\delta^{13}C = -7.6‰$, $\delta^{18}O = 41‰$
Ne	0.0018	
Xe	0.0005	
CH_4^a	0.0002	$\delta^{13}C = -47‰$
SO_2^a	0–0.0001	$\delta^{34}S = 4‰$ (variable)
Kr	0.0001	
H_2	0.00005	$\delta^{2}H = -70 \pm 30‰$
N_2O^a	0.00003	$\delta^{15}N = 15‰$, $\delta^{18}O = 31‰$
CO^a	0.00001	$\delta^{13}C = -27.4‰$, $\delta^{18}O = 24.6‰$
NO_2	0–0.000002	
NH_3^a	0.000001	
O_3^a	0–0.000001	
H_2O_{VAPOUR}	0.001–3	

[a]Anthropogenically sensitive

different heights and analyzed them for their oxygen isotopic composition (Fig. 6.1).

Both $\delta^{18}O$ and $\delta^{17}O$ were measured in CO_2 and it was immediately noticed that the observed extent of fractionation was not as theoretically predicted for mass-dependent fractionation (as described in Chap. 3). This is an interesting notion and one that was already mentioned in Chap. 1, so let us spend some time introducing a new concept.

In a general principle, as noted in Chap. 1, the extent of fractionation—both kinetic and in equilibrium—will be

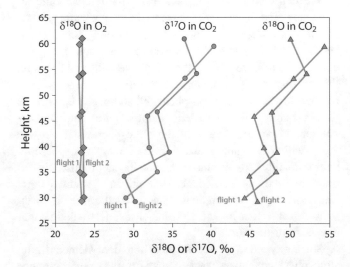

Fig. 6.1 Oxygen isotopic composition for O_2 and CO_2 at different heights in the stratosphere. Both $\delta^{18}O$ and $\delta^{17}O$ were measured in CO_2. Data from [8, 9]

directly proportional to the difference in mass between different isotopes. After all, hydrogen isotopes fractionate 8 times more than oxygen isotopes, as the relative mass difference between ^{2}H and ^{1}H (100%) is 8 times higher than that between ^{18}O and ^{16}O (12.5%). We will call the fractionation the extent of which is *predictable* based on relative mass differences between isotopes, *mass-dependent fractionation*. If, for instance, we measure the three oxygen isotopes (^{16}O, ^{17}O, and ^{18}O), calculate the $\delta^{17}O$ and $\delta^{18}O$, and plot them one against the other, they should be situated on a 1:2 line if we have mass-dependent fractionation (Fig. 6.2). (In reality, the slope varies from 0.500 to 0.526 because of various kinetic effects.)

However, several gas species produced in the upper atmosphere, above the tropopause (approximately 6 km above the poles and 16 km above the tropics), have oxygen isotopes that show a variable degree of *mass-independent fractionation*. This means that the extent of fractionation during their formation is not proportional to the relative mass difference between isotopes (Fig. 6.2). This is a rare situation and the upper atmosphere is one of the few settings where it currently occurs in nature.

When ozone (O_3) was produced in the laboratory by photolysis ($O_2 + O \rightarrow O_3$), starting with $\delta^{17}O = \delta^{18}O = 0‰$, the product ozone and the residual oxygen were situated on a line with slope of 1. This means that the fractionation during this process was purely mass-independent (Fig. 6.2). If we compare this to the naturally occurring upper atmosphere gas species, we will observe that they are

Fig. 6.2 Oxygen isotope
composition, $\delta^{17}O$ versus $\delta^{18}O$,
of various atmosphere gas
species, Bulk Silicate Earth, and
of laboratory-produced ozone.
Gas species below the tropopause
were formed by mass-dependent
fractionation and are situated on a
line with a slope of approximately
0.5 (varying between 0.500 and
0.526 in nature). Species
observed in the upper
atmosphere, above the
tropopause, were formed by at
least some amount of
mass-independent fractionation.
Ozone is produced in the
laboratory by pure
mass-independent fractionation.
Data from [7]

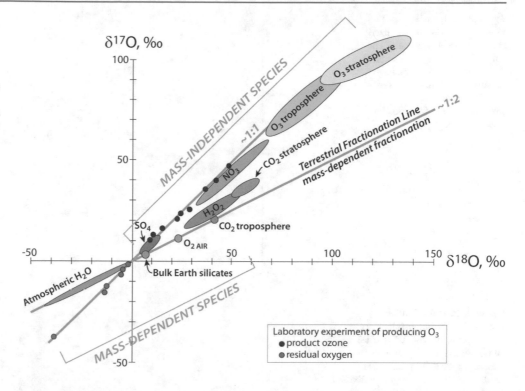

Fig. 6.2 Oxygen isotope composition, $\delta^{17}O$ versus $\delta^{18}O$, of various atmosphere gas species, Bulk Silicate Earth, and of laboratory-produced ozone. Gas species below the tropopause were formed by mass-dependent fractionation and are situated on a line with a slope of approximately 0.5 (varying between 0.500 and 0.526 in nature). Species observed in the upper atmosphere, above the tropopause, were formed by at least some amount of mass-independent fractionation. Ozone is produced in the laboratory by pure mass-independent fractionation. Data from [7]

situated somewhere between pure mass-dependent fractionation (line with a slope of ~1:2) and pure mass-independent fractionation (line with a slope of ~1:1; Fig. 6.2). In other words, the fractionation that produced the naturally existing gas species in the upper atmosphere was partially mass-dependent and partially mass-independent. We can even quantify the proportion of mass-independent fractionation, as the difference between the observed $\delta^{17}O$ and the expected $\delta^{17}O$ if we had only pure mass-dependent fractionation:

$$\Delta^{17}O = \delta^{17}O - 0.5\delta^{18}O$$

A similar situation can be seen with sulphur isotopes. This can be observed when the three main sulphur isotopes (^{32}S, ^{33}S, and ^{34}S) are measured in reduced (sulphides, e.g., pyrite) and oxidized (sulphates, e.g., barite) mineral species from geological samples. If the calculated $\delta^{33}S$ and $\delta^{34}S$ are then plotted against each other, we are able to observe that few samples fall near the mass-dependent fractionation line (Fig. 6.3). The slope of the mass-dependent fractionation line for sulphur isotopes is approximately 0.52 in natural samples.

We will also notice that the sulphides always have positive $\Delta^{33}S$ and that sulphates have a negative $\Delta^{33}S$: as a general rule, sulphides will have higher $\delta^{33}S$ and $\delta^{34}S$, as sulphur is more strongly bound within the sulphides than within the sulphates. The extent of mass-independent

fractionation of an individual sample, $\Delta^{33}S$, is calculated in a similar way as for oxygen isotopes:

$$\Delta^{33}S = \delta^{33}S - 0.52\delta^{34}S$$

We mentioned earlier that the atmosphere interacts with and influences other Earth "spheres": let us see how the geological record can inform us about the evolution of the atmosphere. In this case, we will plot the extent of mass-independent fractionation, $\Delta^{33}S$, against the age of a sulphide or sulphate sample that has been formed at the surface of the Earth and preserved in the geological record (Fig. 6.4).

From the earliest preserved, both reduced and oxidized, mineral species until the appearance of oxygen in Earth's atmosphere at approximately 2.3 Ga, mass-independent fractionation was significant and ranged from −1.5 to 2‰ (Fig. 6.4). As oxygen accumulated in the atmosphere and ozone started forming in the stratosphere, the extent of mass-independent fractionation decreased further, to almost 0. Before the Great Oxygenation Event (also confirmed by other observations in the geological record), the fractionation of sulphur isotopes was dominated by atmospheric reactions, specifically through the photodissociation of SO_2 because of the presence of ultraviolet (UV) radiation on Earth's surface. Samples forming at the surface of the Earth were formed by mass-independent fractionation, meaning that UV rays—originating from the Sun—were present at ground level. However, once oxygen accumulated in the atmosphere and the ozone layer formed, the formation of both reduced and

Fig. 6.3 Sulphur isotopes of reduced (pyrite) and oxidized (barite) sulphur minerals. The sulphides always have positive $\Delta^{33}S$ and the sulphates have a negative $\Delta^{33}S$, but very few samples sit on the mass-dependent fractionation line. Data from [3]

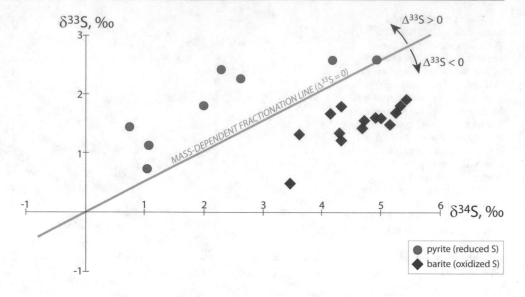

Fig. 6.4 Extent of sulphur isotopes' mass-independent fractionation in reduced (pyrite) and oxidized (barite) sulphur minerals species formed at the Earth's surface. As oxygen started appearing in Earth's atmosphere and ozone started forming, mass-independent fractionation first diminished significantly and then almost disappeared. Data from [4]

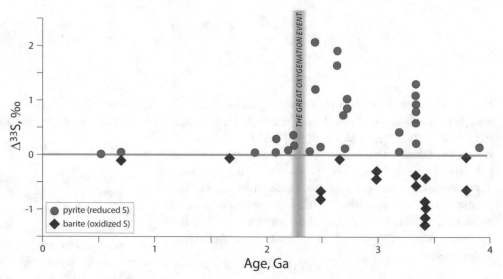

oxidized sulphur minerals at the surface of the Earth was dominated by biologically mediated reduction–oxidation chemical reactions and mass-dependent fractionation. A corollary of this transition is that animals and other life-forms were now able to live on land because of the availability of oxygen, but they will also be protected by harmful UV rays, that were now almost entirely absorbed by the ozone layer.

6.2   Isotopes and Climate Change

Climate change is a central and very important issue today that affects the totality of human societies and, most importantly, the biosphere, the atmosphere, and the hydrosphere. It is extensively studied by scientists, hotly debated by anyone and everyone, and has become a by-word of

political and societal polarization in many countries. However, it doesn't have to be so: the fact of changing atmospheric temperatures is incontrovertible, and evidence that human activity has played a central role in this change cannot be denied. Here, we will review the main causes of climate change, the different processes involved, and, more to the point, how isotopes are helping us to understand changing temperatures in the past and present.

Climate change is not a new discovery: as far back as ancient Greece, philosophers such as Aristotle and his disciple Theophrastus were aware that climate changes. They were even able to propose sound theories as to the cause of the change, some of which hold true today. In more modern times, in the sixteenth–eighteenth centuries, people like Charles de Montesquieu and Jean-Baptiste Dubos of France, David Hume of England, and Thomas Jefferson (the third president of the United States) were commenting on the change in climate and the effects of this change on society. As the scientific revolution properly kicked into gear in the early nineteenth century, scientists John Tyndall and Svante Arrhenius—among others—started speculating about the relationships between the concentrations of gases in the atmosphere (H_2O, CO_2, O_3, CH_4) and the atmosphere's temperature. These scientists were able to lay the foundation of the science behind greenhouse effect and climate change. These efforts then culminated in the first specific and concrete factual evidence of climate change, provided by certain by Guy Callendar.

Guy Callendar was an engineer and inventor, but also an amateur climatologist, by modern terms. He was able to put his hands on an extensive data set of temperature measurements from around the world and compile them to demonstrate that the air temperature was indeed rising at a steeper rate than in the pre-industrial era (Fig. 6.5).

Callendar announced to the world that overall atmospheric temperature had risen by nearly 0.5 °C between 1890 and 1935. He also claimed that human emissions of CO_2 from burning fossil fuel (mostly coal) could cause a "greenhouse effect" and that this was the cause of the warming. Building on Arrhenius' work, Callendar asserted that increasing CO_2 concentration in the atmosphere resulted in better retention of thermal energy and thus caused increasing temperatures. This theory was called the Callendar Effect and its predictions were later proven to have been remarkably accurate, when climate science took off in the 1950s and 1960s.

Callendar did not leave his speculations there: a few years later, in 1941, he published another paper explaining exactly how the increase of CO_2 concentrations—and also those of other "greenhouse" gases such as N_2O and CH_4—caused an increase in temperature (Fig. 6.6).

Calendar focused on the increase of "greenhouse" gases as the cause of rising atmospheric temperatures; however, there are yet other causes at work. On the planetary scale and in the long term, there are two main groups of causes for temperature variations on the Earth's surface, and these are generally grouped as astronomical and terrestrial. The former is controlled by the slight variations of Earth's orbit ellipticity and rotational axis inclination; they entirely predictable (according to the Milankovitch cycles) and have periods of 19, 22, 24 thousand years (for the precession), 41 thousand years (for the obliquity), and 95, 125, and 400 thousand years (for the eccentricity). To these we can add the variations in solar activity, which are less cyclic and thus less predictable.

The terrestrial causes are more complex and even less predictable, as they depend on two forcing feedbacks working in opposite directions (Fig. 6.7). The first is called the albedo feedback: as we have more snow and ice on Earth's surface, more and more of Sun's heat will be reflected (high albedo), causing further cooling of the atmosphere. It is when this feedback had reached extreme proportions that the Earth resulted in being covered in ice from pole to pole (the "Snowball Earth" episodes). The second is called water vapour feedback: when air temperatures increase, more evaporation occurs resulting in elevated greenhouse effect and higher air temperatures. If this feedback is left alone, it can get carried away and result in high air temperatures with no ice, including on the poles. Indeed, more than half of Earth's history was ice-free, as is demonstrated by the sedimentary record.

Let us further complicate things by adding the moderating, or stabilizing feedback, which involves erosion and the weathering of rocks on the continents, an increase of CO_2 in the atmosphere, and greenhouse effect, and the tectonic

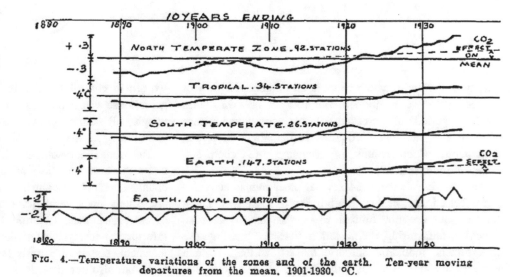

Fig. 6.5 Compilation of temperature measurements from around the globe, compiled by Guy Callendar and published in 1938 [1]. Notice the "CO_2 effect" on the overall Earth averages line

FIG. 4.—Temperature variations of the zones and of the earth. Ten-year moving departures from the mean, 1901–1930, °C.

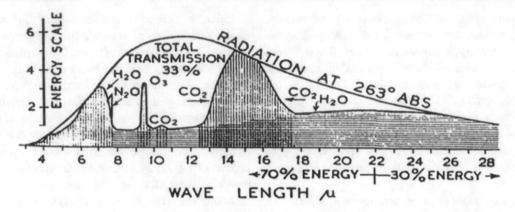

Fig. 6.6 A diagram from Guy [2] paper, depicting the absorption spectra of different gases in the atmosphere, for the infrared portion of the energy spectrum. Water, methane, carbon dioxide, nitrous dioxide, and ozone are included: these are the same main "greenhouse" gases we consider today

Fig. 6.7 Schematic representation of the three main forcing feedbacks affecting Earth's climate. These do include the tectonic cycle, with periods of increased volcanic activity and thus increased output of CO_2 and other volatiles in the atmosphere, and of subduction, which recycles carbon in Earth's mantle

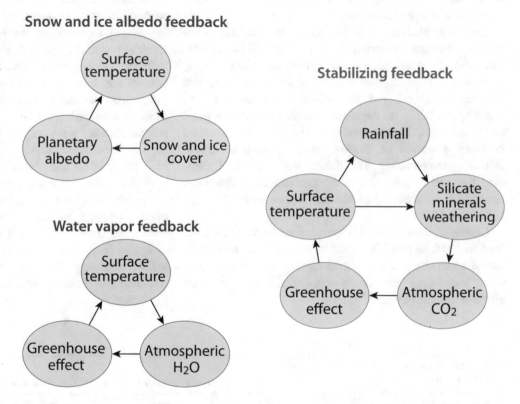

activity of the Earth, which is not at all influenced by any of the feedbacks we mentioned, but follows its own cycles involving periods of increased volcanic activity and release of more CO_2 and other volatiles in the atmosphere, and subduction, which results in recycling of carbon and decreasing temperatures.

And, of course, there are several other human activities that affect climate, besides the burning of fossil fuel. These include—but are not limited to—emissions of aerosols, cement manufacturing, deforestation, livestock breeding, use of fertilizers, industrial activity; the list is long. Human activities are, however, the only climate change factor that we are in control of. The bad news is that the effects of greenhouse gas emissions on climate change heavily outweigh all other causes of atmospheric temperature variations.

But what about isotopes, I hear you ask? Isotopes happen to be the best and most reliable indicator of past climate changes, including changes in the actual temperatures of the air. There are several different methods that are used to access past climate variations, and these include biological information (analysis of pollen and beetle distribution, for instance), historical records (such as Callendar's compilation), but also tree records (the information recorded in tree rings, each corresponding to a year), the glacial record (studying ancient ice), and the geological record.

The change in CO_2 concentration in the air has been demonstrated to be directly related to the carbon isotopic signature (Fig. 6.8), and the change in CO_2 has been directly related to temperature variations.

Thus, if we have access to ancient air and are able to measure its CO_2 concentrations and carbon isotopic composition of its CO_2, we will have a very good record of air temperature variations. But where can we find ancient air? There is only one place where we can obtain this, and that is within ancient ice, and there is plenty of it in the two polar regions. Ancient ice contains tiny bubbles of air trapped during ice formation. Today, scientists take drill cores of ice, determine their age, and extract the ancient air, providing a very powerful proxy for paleoclimate variations (Fig. 6.9).

There is another powerful proxy for tracking paleoclimate variations: oxygen isotopic record in sedimentary rocks. Let us start by considering a situation of globally low temperatures; a glacial period. There will be a higher amount of ice accumulated in the polar regions, leading to a higher degree of sequestration of ^{18}O (and a lower degree of available ^{18}O). This occurs because ice is more likely to contain ^{18}O than water, oxygen being more strongly bound in ice than in water (see Chap. 1). In other words, the remaining free water (liquid and vapour) will have lower $\delta^{18}O$ values during glacial periods. That water—be it in the ocean, the atmosphere, in lakes and rivers, or in ice—will have a clearly lower isotopic composition than water during warm, interglacial periods.

In other words, there is a clear relationship between water oxygen (and hydrogen) isotopic composition and global air temperature: warmer, interglacial conditions are reflected in higher $\delta^{18}O$ values of water. Importantly, the isotopic composition of water is recorded in the geological record via the formation of minerals in the ocean or on the continents. This can be seen in minerals such as biogenic carbonates (foraminifera, molluscs, ostracodes, ooliths, stromatolites, etc.), biogenic phosphates (teeth or bone), silica (cherts, diatoms), alunite, biogenic cellulose, or clay minerals (kaolinite and smectite). As we noticed in Chap. 1, the fractionation factor, α, between water and a mineral is a function of temperature. Thus, if we analyze the oxygen isotopic composition of a mineral preserved in sedimentary rocks, we can approximate the mean annual air temperature at the time of precipitation of this mineral.

In practice, this is a challenging proposition, and several difficulties appear very quickly. Firstly, estimating the fractionation factor of stable isotopes between any phase and water becomes particularly challenging at low temperatures. Secondly, we have to know, or assume, the water isotopic composition. Thirdly, the presence of kinetic effects—particularly in biogenic minerals—can obscure the fractionation factor. And finally, diagenetic effects (changes to the sediments as they are exposed to increased temperature and pressure during burial) can also affect the isotopic signature of the minerals analyzed.

Let us take, for instance, the estimation of the fractionation factor between carbonates and water as a function of temperature. It is a linear relationship:

$$\delta^{18}O_{mineral} - \delta^{18}O_{water} \sim 1000 \ln \alpha = a/T + b$$

where T is the temperature in degrees Kelvin. The two descriptors of the linear relationship, a and b, vary significantly in the published literature: a varies from 18.07×10^3 to 18.56×10^3, and b varies from 31.08 to 33.49 (this corresponds to relative differences of approximately 7%). In practical terms, this translates to differences of a few degrees Celsius in the estimated water temperature. The situation is similar to the silica–water fractionations, where different published fractionation factors can result in differences of up to 10 °C, which is not ideal. The situation is much better with carbonates–water fractionation for planktonic foraminifera, which is described by the equation:

$$T(°C) = 16.9 - 4.2(\delta^{18}O_C - \delta^{18}O_W) + 0.13(\delta^{18}O_C - \delta^{18}O_W)^2$$

This equation can be summarized to say that for each degree Celsius of temperature decrease, the $\delta^{18}O$ in the shell augments by approximately 0.25‰. In other words,

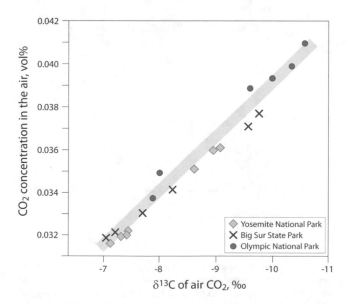

Fig. 6.8 Relationship between CO_2 concentration in the air and the isotopic composition of air CO_2. This corresponds to a 0.02‰ change in $\delta^{13}C$ for every ppm$_V$ change in CO_2 concentration. Data from [5]

Fig. 6.9 Temperature record in
the Vostok ice core, Antarctica,
showing a strong correlation to
CO_2 concentrations [6]

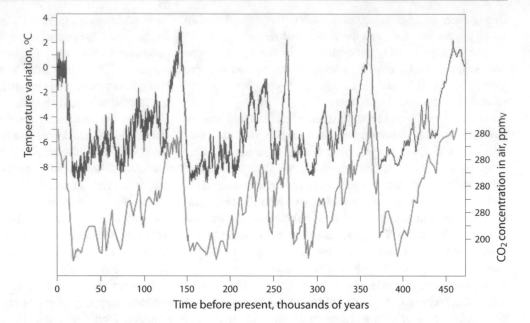

Fig. 6.9 Temperature record in the Vostok ice core, Antarctica, showing a strong correlation to CO_2 concentrations [6]

foraminifera are faithful witnesses to changes in water temperature, as seen in Fig. 6.10.

After compiling oxygen isotope data in the sedimentary record, from benthic foraminifera, as far back in time as possible, scientists noticed an interesting thing: extreme negative excursions of the $\delta^{18}O$ values were occurring. These were interpreted as reflecting strong dips in the global mean atmospheric temperature at the Earth' surface. In turn, these low temperatures resulted in a strong increase in the extent of continental ice sheets, which reached much closer to the equator. In extreme cases, the totality of the Earth's surface was covered by ice, which is why these episodes were called the Snowball Earth. This likely occurred at least twice in Earth's history, at around 650 and 700 million years ago. This has been confirmed by the sedimentary record: glacial sedimentary deposits have been found in equatorial regions. Contrasting to the Snowball Earth is an ice-free Earth, which was the condition of the Earth's surface for more than half of its history.

Importantly, because of the effects of the difference feedback loops mentioned above—in particular, the moderating one—the sedimentary record will reflect the variations in temperature. Low global air temperatures will correspond to low $\delta^{18}O$ and low $\delta^{13}C$ values of carbonate layers; this is also true in lower $^{87}Sr/^{87}Sr$ ratios, which correspond to lower weathering rates, as the availability of free running water on the continents becomes very low.

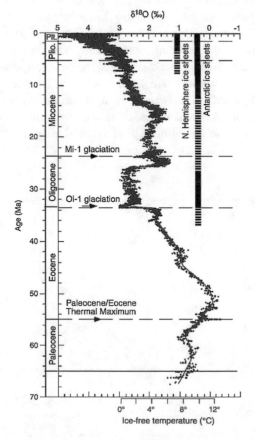

Fig. 6.10 Oxygen isotopes in foraminifera, reflecting significant changes in ocean temperatures over the last 70 million years. Sharp drops in $\delta^{18}O$ correspond to the presence of ice sheets in polar regions; while specific glaciation events can be identified on the basis of oxygen isotopes. From [10], modified

Further Reading

Unfortunately, for some reason, there are very few books that exist specifically on the application of isotopes in the study of the atmosphere. Rather, the topic is covered as parts of other books, and here is a short selection:

Isotopes in the Earth Sciences, H.-G. Attendorn and R. Bowen, Springer Science, 1988, ISBN 978-041-253710-3. The relevant chapter is 9, Environmental isotopes in the atmosphere and hydrosphere.

Stable Isotope Geochemistry, J. Hoefs, Springer-Verlag, 1997, ISBN 3-540-61126-6. The only chapter of interest for this topic is 3.9 Atmosphere.

Stable Isotopes and Biosphere—Atmosphere Interactions: Processes and Biological Controls, L.B. Flanagan, J.R. Ehleringer, and D.E. Pataki, Editors, Elsevier, 2004, ISBN 978-008-052528-0. Most relevant are Chaps. 15 and 16.

Helium Isotopes in Nature, B.A. Mamyrin and I.N. Tolstikhin, Elsevier, 2013, ISBN 978-148-328980-9. Chapter 10 is the most relevant.

Questions

- What is the current atmosphere made of? When and how was it formed? How many atmospheres were there before and what gases were they dominated by?
- What is the vertical structure of the atmosphere and how does it affect the isotopic composition of its constituent gases?
- What is mass-independent fractionation? Where does it occur and why? How can we detect and quantify it? The isotopes of which elements are typically affected by it? How can we detect in the rock record?
- How, and in what way, can isotopes inform us about past climate changes? What isotopes and what sort of material do we use for that? How can we calculate paleotemperatures?

References

1. Callendar GS (1938) The artificial production of carbon dioxide and its influence on temperature. Q J R Meteorol Soc 64:223–240
2. Callendar GS (1941) Infra-red absorption by carbon dioxide, with special reference to atmospheric radiation. Q J R Meteorol Soc 67:263–275
3. Farquhar J, Savarino J, Jackson TL, Thiemens MH (2000) Evidence of atmospheric sulphur in the martian regolith from sulphur isotopes in meteorites. Nature 404:50–52
4. Golding SD, Duck LJ, Young E, Baublys KA, Glikson M, Kamber BS (2011) Earliest seafloor hydrothermal systems on earth: comparison with modern analogues. In: Golding SD, Glikson M (eds) Earliest life on earth: habitats, Environments. Springer Science, Berlin, pp 15–49
5. Keeling CD (1958) The concentration and isotopic abundance of atmospheric carbon dioxide in rural areas. Geochim Cosmochim Acta 13:322–334
6. Petit JR, Jouzel J, Raynaud D, Barkov NI, Barnola J-M, Basile I, Bender M, Chappellaz J, Davis M, Delaygue G, Delmotte M, Kotlyakov VM, Legrand M, Lipenkov VY, Lorius C, Pepin L, Ritz C, Saltzman E, Stievenard M (1999) Climate and atmospheric history of the past 420,000 years from the Vostok ice core, Antarctica. Nature 399:429–436
7. Thiemens MH (2006) History and applications of mass-independent isotopes effects. Annu Rev Earth Planet Sci 34:217–262
8. Thiemens MH, Jackson TL (1995) Observations of mass-independent oxygen isotopic composition in terrestrial stratospheric CO_2, the link to ozone chemistry, and the possible occurrence of Martian atmosphere. Geophys Res Lett 22:255–257
9. Thiemens MH, Heidenreich JE (1983) The mass-independent fractionation of oxygen: a novel isotope effect and its possible cosmochemical implications. Science 219:1073–1075
10. Zachos J, Pagani M, Sloan L, Thomas E, Billups K (2001) Trends, rhythms, and aberrations in global climate 65 Ma to present. Science 292:686–693

7.1 Isotopes in the Human Body

The human body is an exceptionally complex assemblage of tissues, organs, and fluids, all working in perfect harmony and balance. Harmony and balance may be disrupted by injury and illness, and an army of health professionals are working hard to restore them and keep us all in good health. Some of them are deeply engaged in a better understanding of how the human body works, and they often use isotopes to help them, as all processes occurring in the body will cause measurable isotopic fractionations. In this chapter, we will examine how isotopes can assist us in the pursuit of better health and better understanding of the workings of the human body.

One way to look at the human body is by its chemical and isotopic composition. The average weight of a human is 70, 25 kg of which is bone and connective tissue. Water represents around 72% of the lean mass, or some 30–35 kg. Three elements, oxygen, carbon, and hydrogen, make up 93% of the average body mass, with the rest being nitrogen, calcium, phosphorus, potassium, sulphur, sodium, chlorine, magnesium, and several trace elements (Table 7.1). When the average natural abundances of the main stable isotopes of these elements are considered, we can have a very good idea of how much of each isotope is present in the average human body (Table 7.1). For instance, the average human contains 45.39 kg of ^{16}O, 93.3 g of ^{18}O, 12.8 kg of ^{12}C, 138.6 g of ^{13}C, etc. Even minor isotopes of minor elements will be present in measurable amounts, such as ^{37}Cl, of which there is 33.9 g in the average human body. In other words, we have several isotopic systems to choose from when studying the functioning of the human body.

The applications of isotopes to human health are numerous and we cannot describe them all here. Only a few selected examples will be given to represent typical and common applications. However, a few commonalities are present and we will mention them first.

7.2 Fractionation in the Human Body

Fractionation of stable isotopes in a living organism is always a kinetic process, i.e., it does not occur in equilibrium conditions and all life-related processes and reactions are non-reversible. The major factor controlling the extent of fractionation in kinetic processes is the difference in diffusion rates—due to differences in mass—between isotopes and molecules that are made of different isotopes. One example is breathing: oxygen is inhaled and taken into the lungs where it incorporates into the bloodstream of the body, while CO_2 is exhaled. It is estimated that the lighter $^{16}O_2$ molecule will diffuse approximately 3% faster than the heavier $^{16}O^{18}O$ molecule, which leads to metabolizing $^{16}O_2$ 1.3% faster than $^{16}O^{18}O$. O_2 diffusion in the lungs and utilization in tissues will thus lead to oxygen isotopic fractionation; however, ventilation (breathing) and transportation in the bloodstream do not lead to any fractionation. It can be considered that breathing consists of a series of fractionating and non-fractionating processes, with an overall fractionation effect on oxygen in which $^{16}O_2$ is transported approximately 0.8% faster than $^{16}O^{18}O$ (when the person is at rest).

7.3 You Are What You Eat

Fractionation of several isotopic systems in the human body is subject to two competing phenomena, also mentioned in Chap. 5:

(1) Preferential ingestion and metabolization of the lighter isotopes. Molecules containing lighter isotopes are less strongly bound within the ingested nutrition source; therefore, the body will expend less energy digesting these molecules than digesting molecules made of heavier isotopes. It is thus energetically more efficient

P. Alexandre, *Isotopes and the Natural Environment*, Springer Textbooks in Earth Sciences,
Geography and Environment, https://doi.org/10.1007/978-3-030-33652-3_7

Table 7.1 Amount, in kilograms, and proportions, in percent, of the elements and their isotopes that make up the average human body, assuming a weight of 70 kg. Two elements, phosphorus and sodium, have only one isotope each

Elements	% of element	Weight of element (kg)	Major isotopes	% of isotope	Weight of isotope (kg)
O	65	45.50	^{16}O	99.757	45.389
			^{18}O	0.205	0.093
C	18.5	12.95	^{12}C	98.93	12.811
			^{13}C	1.07	0.139
H	9.5	6.65	^{1}H	99.9885	6.649
			^{2}H	0.0115	0.001
N	3.2	2.24	^{14}N	99.632	2.232
			^{15}N	0.368	0.008
Ca	1.5	1.05	^{40}Ca	96.941	1.018
			^{44}Ca	2.086	0.022
P	*1.0*	*0.70*	^{31}P	*100*	*0.700*
K	0.4	0.28	^{39}K	93.2581	0.261
			^{41}K	6.7302	0.019
S	0.3	0.21	^{32}S	94.93	0.199
			^{34}S	4.29	0.009
Na	*0.2*	*0.14*	^{23}Na	*100*	*0.140*
Cl	0.2	0.14	^{35}Cl	75.78	0.106
			^{37}Cl	24.22	0.034
Mg	0.1	0.07	^{24}Mg	78.99	0.055
			^{25}Mg	10.00	0.007
			^{26}Mg	11.01	0.008
Trace elements	*0.1*	*0.07*			
Total	**100**	**70**			

for the body to ingest and metabolize lighter isotopes. In addition, lighter isotopes will diffuse faster and amplify the same preference for lighter isotopes.

(2) Preferential elimination of the lighter isotopes from the body. Heavier isotopes will be more strongly bound in the tissues of the body and will tend to stay behind in the body. This effect is amplified by their lower diffusion rates.

The overall result of the interplay between these two phenomena is the slight overall enrichment in heavy isotopes.

A major factor affecting the isotopic composition of the human body is diet, most noticeable with carbon and nitrogen isotopes. It has been demonstrated by numerous studies that body tissue will be enriched in ^{13}C and ^{15}N in people eating more meat (higher trophic level food) than those eating less, and even more enriched than vegetarians and vegans, who prefer food from lower trophic levels (Fig. 7.1). This effect, the trophic shift that we examined in Chap. 5, affects the totality of the food web, not only people. In people, the diet-related—or trophic shift-related—

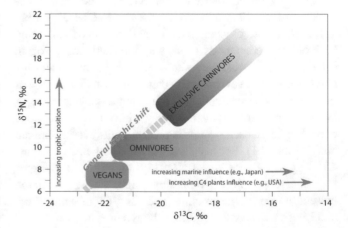

Fig. 7.1 Trophic shift, illustrated by carbon and nitrogen isotopic compositions, measured in hair or fingernails (thus representing the diet from the last few months). Data from [1, 2], and the author's data

difference between vegans and omnivores has been estimated at somewhere between 4 and 9‰ for ^{15}N and ^{13}C, with significant variations between individual people and populations.

7.4 Measuring the Total Energy Expenditure

An important health index is the Total Energy Expenditure (TEE), which is the amount of calories burned by the human body in 1 day, adjusted to the amount of activity (sedentary, moderate, or strenuous). It varies significantly from person to person and can be derived from a person's age, gender, weight, and activity level. On the other hand, the Physical Activity Level (PAL) is defined as the ratio between the Total Energy Expenditure and the Resting Energy Expenditure (REE): PAL = TEE/REE. The Physical Activity Level for an average sedentary person is approximately 1.5, meaning that their everyday physical activity requires 50% more energy than their basal metabolism. With age and increased sedentary habits, PAL will decrease to ca. 1.3, whereas it will be above 2 for an endurance athlete (Fig. 7.2). In other words, the basal metabolism during rest of those two extreme cases may require approximately the same daily calorie uptake. For example, a person at rest may require approximately 2000 ccal, but a sedentary elderly person will require only approximately 600 extra calories during normal daily activity, whereas the elite athlete will consume at least another 2000 ccal, necessary for their training effort.

Two considerations underline the importance of precisely measuring the Total Energy Expenditure:

(1) Any dietary calorie uptake in excess of the Total Energy Expenditure will result in fat as an energy store being laid down, while if the TEE is higher than the dietary calorie intake, fat will be oxidized to release extra energy. This has direct implications for healthy body weight, most importantly in cases regarding co-existing pathologies.

(2) Any Physical Activity Level below 1.2 is considered dangerously low and can lead to severe complications and death. This is particularly true for the elderly and frail patients.

Because of these considerations, measuring the Physical Activity Level is very important and is often required in specific pathological situations. As PAL the ratio of Total Energy Expenditure and the Resting Energy Expenditure, we need to reliably measure these two amounts. The latter one, REE, is measured by a method called indirect calorimetry, which consists of measuring the production of CO_2 and nitrogen waste (such as urea, for example) and the consumption of oxygen. Typically, the patient, maintained at rest, is placed in an impermeable canopy or suit and the amounts of O_2 uptake and CO_2 production are measured. The idea here is that higher O_2 uptake and CO_2 production relate to higher REE: by using the exact amounts and accounting for age, gender, and weight, the Basal Metabolic Rate and thus the REE are calculated.

This is a very well-established and well-understood technology that dates back two centuries (even though it has been heavily employed by medical professionals and elite sport scientists only in the last two decades or so). However, how can we measure the Total Energy Expenditure, though, without requiring the patient to go about their

Fig. 7.2 Two extreme cases: an elderly person with reduced physical activity and an endurance athlete. They may have similar resting energy expenditure, but their physical activity levels will be radically different

daily life, and never mind perform strenuous tasks, lugging around heavy equipment not dissimilar to a spacesuit? Well, this is where isotopes come to the rescue.

First, the patient is made to drink isotopically labelled water: this is water that has a specific and unique isotopic composition that is very different from that of the human body. In this application, the water has been isotopically labelled for both hydrogen and oxygen. As water participates in the body's metabolic processes, oxygen will equilibrate isotopically with the CO_2 that the body eliminates through breathing, while hydrogen will be eliminated only as water, mostly through urine (and perspiration). Both CO_2 and urine are analyzed every few hours, during several days, for O and H isotopes and the decreases of $\delta^{18}O$ and δ^2H are monitored as they decrease to typical body values (Fig. 7.3). Importantly, the elimination rate of the two tracers is not the same: CO_2 production depends on physical activity, but peeing does not. Thus, it is the difference in elimination rates between the two tracers that is used to calculate quite precisely the CO_2 output and hence the TEE, using the empirical formula

$$r_{CO2} = 0.455 \times TBW \times (1.01 K_O - 1.04\ K_H)$$

where r_{CO2} is the CO_2 production rate, TBW is the total body weight, and K_O and K_H are the oxygen and hydrogen turnover rates that have been calculated from the slopes of the elimination plots (Fig. 7.3). The CO_2 production rate is then directly related to TEE.

Importantly, this method does not rely on naturally occurring isotopes, but uses isotopes as markers. Indeed, there are several other methods using stable isotopes as labels and markers and we will consider another one right away.

7.5 Helicobacter Pylori and Ulcer

In the early evening of 10 December 2005, two scientists, Barry Marshall and Robin Warren, were happily enjoying some excellent champagne. They had a good reason to celebrate: they had just been awarded the Nobel Prize in Medicine "for their discovery of the bacterium *Helicobacter pylori* and its role in gastritis and peptic ulcer disease", more than 20 years prior. Ulcer is one of those diseases that make your life totally and utterly miserable for a very long time, severely affecting your quality of life, but rarely have the ability to kill you. It was not known, before Marshall and Warren's discovery, what causes ulcer or how to treat it properly. All sorts of remedies were available, some just short of witchcraft and others with a solid veneer of science, but all were mostly ineffective.

H. Pylori (Fig. 7.4) is a very interesting beast: it thrives where it should not, in the harsh and strongly acidic conditions of the stomach. (It is also responsible for more than 90% of duodenal ulcers and up to 80% of gastric ulcers, which makes it hated by very many sufferers including the author of these lines.) *H. Pylori* manages to survive in gastric acid by excreting large amounts of a particular enzyme called urease. This enzyme breaks down any urea present in the stomach to carbon dioxide and ammonia:

$$CON_2H_4 + H_2O \rightarrow CO_2 + 2NH_3$$

The ammonia thus produced neutralizes any acid found directly in the vicinity of the bacteria, allowing the bacteria to survive (Fig. 7.5). So far, so good. However, ammonia is

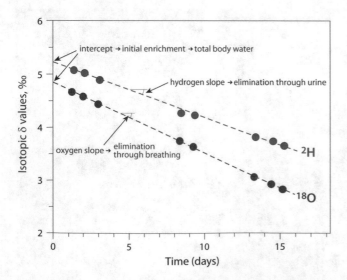

Fig. 7.3 The metabolic rate is estimated from the difference between the elimination speeds of oxygen and hydrogen, reflecting the CO_2 turnover rate. The isotopic compositions of both elements are measured in urine. From [3], modified

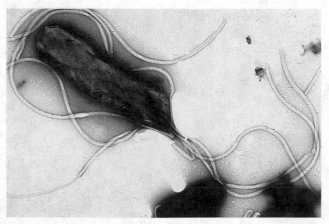

Fig. 7.4 Electron micrograph of *Helicobacter Pylori* with multiple flagella

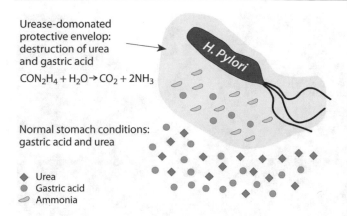

Fig. 7.5 Simplified schematic representation of *H. Pylori*'s protective mechanism: it releases the enzyme urease, breaking down gastric acid and urea to ammonia and CO_2

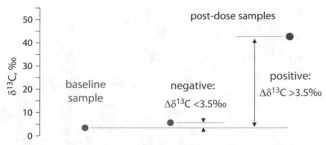

Fig. 7.6 Interpretation of the *H. Pylori* breath test: any sample with $\Delta\delta^{13}C$ larger than 3.5 is considered positive; higher values are considered to reflect a higher degree of infection

toxic to us and our body is constantly producing urea to neutralize ammonia. Thus, the presence of *H. Pylori* in the stomach has a two-fold negative effect: it constantly produces a toxic substance that the body constantly has to fight, and it reduces the amount of gastric acid required for digestion of our food, thus forcing the body to produce more. The acid production is often excessive and commonly leads to lesions in the stomach or duodenum lining, also known as ulcers.

Well, it didn't take long for scientists to develop a reliable diagnostic tool: they noticed that the "protection mechanism" of *H. Pylori* involves the production of CO_2, which is absorbed into the bloodstream and then released from the lungs when exhaling. The "Urea Breath Test", or just Breath Test, is based on this reaction, and it works like this: Firstly, the patient is given an acidic drink, such as a glass of orange juice, to quickly close the duodenal sphincter and thus contain the stomach contents. After that, a baseline breath sample is taken. Then, the patient is given a (tasteless and odourless) drink containing ^{13}C-enriched urea. The *H. Pylori*, on arrival of a large amount of urea, quickly starts to break it down, as described above, and release CO_2: this CO_2 is enriched in ^{13}C from the spiked urea drink. After half an hour, the patient gives a second breath test. The pre-drink and post-drink breath samples are analyzed for ^{13}C and the results are compared: the larger the number of *H. Pylori* present, the larger the post-drink $\delta^{13}C$ will be and the larger the difference between the pre-drink and the post-drink ($\Delta\delta^{13}C$) will be (Fig. 7.6).

Differences of less than 3.5‰ are considered negative (they are too close to the analytical uncertainty), and those above 3.5% are considered positive, with levels of 30–40‰ over the baseline typical of *H. Pylori* infections. Importantly, the higher the $\Delta\delta^{13}C$, the greater the extent of the infection, giving a quantitative measure of the gravity of the infection. (This author's $\Delta\delta^{13}C$ was on the order of 60%...)

Significantly, this method boasts 96% sensitivity (i.e., only 4% of tests give false positives or false negatives) and 100% specificity: no other infection will produce ^{13}C-rich breath when the patient drinks ^{13}C-enriched urea.

7.6 Don't Cheat! We Will Catch You

Another story, another hero: this time we will consider a hero turned anti-hero, who then perhaps turned back into a hero. Floyd Landis, a faithful lieutenant of a certain Lance Armstrong, struggled with injuries in his cycling career, but managed to do very well in several competitions. He seems to have learned more than one trick from his mentor and master Armstrong: after the latter retired, Landis won convincingly the most arduous and demanding cycling competition in the world, the *Tour de France* (Fig. 7.7), in 2006. Impressively, after struggling on Stage 16 (he finished that stage 23, several minutes from the stage winner and dropped down to 11 in the overall standings), he rode stunningly on Stage 17 (July 20, a brutal mountain stage in the majestic Alps) to come well ahead of reputed "climbers" by nearly 6 min. He jumped back to 3rd overall, and went on to eventually win the competition. That evening, many competitors, journalists, and commentators were asking, how was this possible? Well, the answer is simple: while peacefully giving interviews in the evening before Stage 17, Landis had a testosterone patch stuck on his crotch, a strictly forbidden, but alas widely used and highly effective technique of introducing performance-enhancing drugs into the body. On 27 July 2006, just a few days after the competition ended, his team, Phonak Cycling, announced that a urine sample submitted by Landis on the evening of July 20—the evening after Stage 17!—tested positive for a high ratio of the testosterone to the epitestosterone (T/E ratio). Mayhem ensued and Floyd Landis was eventually stripped of his title and banned for 2 years from any competition. Let us consider how isotopes helped to catch the cheater.

Testosterone ($C_{19}H_{28}O_2$; Fig. 7.8) is a steroid hormone from the androgen group. It has many beneficial anabolic

Fig. 7.7 The Tour De France has a long history of cheating and doping, from its inception to today, in spite of the efforts of the organizers and the International Cycling Union (credit Flickr)

effects, such as helping with the growth of muscle mass and strength, increased bone density and strength, and stimulation of linear growth and bone maturation. In athletics, it's often used because it enhances muscle development, strength, and endurance, and increases the muscles' protein synthesis, which all results in muscle fibres becoming larger and repairing faster than in an average person. Epitestosterone, on the other hand, has the exact same chemistry and nearly the same molecular structure as testosterone, with the only variation being slight differences in molecule endings; it is an inactive epimer of testosterone. It is produced

independently and does not have any of the effects of testosterone; indeed it has no known effects. In an average person, the two are produced approximately in the same amounts, with a testosterone/epitestosterone ratio of approximately 1.2.

If an athlete takes testosterone, this ratio will increase; T/E ratios of up to 4 are tolerated by anti-doping authorities, as it is considered that some natural increase in testosterone can be present, and also to account for analytical uncertainties. In Floyd Landis' urine, 30 min after the end of Stage 17, the T/E ratio was approximately 11! There was no doubt in peoples' mind that he had taken testosterone before Stage 17, but he denied it vigorously. Well, he must have not known about stable isotopes, or he would not have been so vocal in claiming innocence. The standard procedure by the anti-doping agencies worldwide is the following: if the first sample (sample A) is positive, the second sample (sample B) is reanalyzed for its chemical composition, but this time it is also analyzed for its isotopic composition. Testosterone was isolated and analyzed for stable isotopes, and its $\delta^{13}C$ isotopic composition (or "signature") was widely different from all other compounds in his urine, including epitestosterone, which clearly indicated that it was a foreign substance taken artificially.

This is just one of many examples of stable isotopes used forensically. Indeed, most law-enforcement agencies in the world have their own stable isotope laboratories tracing the origin of many substances, mostly organic matter. Stable,

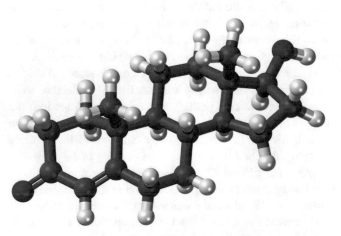

Fig. 7.8 Schematic representation of the testosterone molecule, a steroid and the main hormone responsible for the growth of male primary and secondary sex characteristics

but also radiogenic, isotopes are an extremely powerful authentication method in cases of forgery, for instance, as we will examine in some detail in Chap. 9.

And Floyd Landis in all that?… He took a long look at himself and ended up advocating against organized doping in sports.

Further Reading

There is no shortage of books and journal articles on the topic of isotopes and humans, and here is a brief selection:

Radioisotopes in the Human Body: Physical and Biological Aspects, F. W. Spiers, Elsevier, 2013, ISBN 978-148-325855-3.

Advances in Isotope Methods for the Analysis of Trace Elements in Man Modern Nutrition, M. Jackson and N. Lowe, Editors, CRC Press, 2000, ISBN 978-142-003671-8.

Stable Isotopes in Human Nutrition: Laboratory Methods and Research Applications, S. A. Abrams and W. W. Wong, Editors, CABI, 2003, ISBN 978-085-199796-4.
Bol, R., and Pflieger, C. (2002) Stable isotope analyses (^{13}C, ^{15}N, and ^{34}S) analysis of the hair of modern humans and their domestic animals. Rapid Communications in Mass Spectrometry, 16, 2195–2200.
O'Connell, T.C., Hedges, R.E.M., Healey, M.A., and Simpson, A.H.R.W. (2001) Isotopic composition of hair, nail and bone: modern analyses. Journal of Archeological Science, 28, 1247–1255.

O'Connell, T.C., and Hedges (1999) Investigations into the effect of diet on modern human hair isotopic values. American Journal of Physical Anthropology, 108, 409–425.

Questions

- What is the human body made of, chemically and isotopically speaking?
- What sort of fractionations occur in the human body?
- "You are what you eat": discuss.
- Why is the Total Energy Expenditure important and how do we measure it? The isotopes of which elements are used and how does it work?
- What is the common feature between the isotopic measurement of Total Energy Expenditure and the detection of Helicobacter Pylori, as well as other applications of isotopes to human health?
- How did Floyd Landis get caught? Could he have disguised his use of performance-enhancing drugs?

References

1. Buchardt B, Bunch V, Helin P (2007) Fingernail and diet: stable isotope signatures of a marine hunting community from modern Uummannaq, Northern Greenland. Chem Geol 244:316–329
2. Macko SA, Engel MH, Andrusevich V, Lubec G, O'Connell TC, Hedges REM (1999) Documenting the diet in ancient human populations through stable isotope analysis of hair. Philos Trans R Soc Lond B 354:65–76
3. Murgatroyd PR, Coward WA (2003) Measurement of energy expenditure. In: Caballero B (ed) Encyclopedia of food sciences and nutrition, 2nd edn. Elsevier Science, pp 2098–2103 (2003)

Archaeometry and Society

8.1 What Is Archaeometry?

Archaeology is a well-established science, with its own methods and practices. One branch of archaeology is specifically concerned with measurements applied to archaeology (hence the "-metry" in the name). Both quantitative and qualitative measurements are involved and can be related to

- Identification of materials by chemistry, optics, and mineralogy
- Prospection: the research of where and what to excavate
- Chronometry: dating art and architectural and archaeological artefacts
- Provenance: the search for the site where an artefact was manufactured
- Environmentology: the study of ancient climate, landscape, and geological changes in a given area at a given time
- Biosphere: the research into paleo-ethnic-botany and zoo- and anthropo-archaeology
- Restoration–conservation: how to restore and conserve ancient artefacts.

The methods that are mostly used are chemical (defining the chemical composition of artefacts such as pigments, metal, paper, ceramics, bones, and glass.), mineralogical (defining the mineral composition of artefacts such as ceramics and pigments.), and isotopic (dating or artefacts, studying their provenance, and studying people's ways of life). Here, we will consider a few brief examples of the use of isotopes in archaeometry, as, necessarily, we cannot devote too much space to this otherwise fascinating topic.

8.2 Stonehenge and Population Migration

There is no need to introduce the mystical, unique, enigmatic Stonehenge site. It is the largest and best known prehistoric structure of many found in the British Isles and northwestern France and its purpose was most likely spiritual and religious. Whatever the function of the site and these structures (some of them have been repurposed repeatedly), there is no doubt of the deep cultural and spiritual significance of Stonehenge and similar sites for the people of these times. These people are known as the Beaker people, living during the Bronze Age (approximately 2500–2500 years BC) in what is today South West England.

Well preserved remains of people were found at three separate burial sites in the near vicinity, including two adults in the actual ditch surrounding Stonehenge (Fig. 8.1; [3]). Apart from their apparent age, deduced from their skeletons, little was known of these people, including the knowledge of where they came from. However, radiogenic and stable isotopes are extremely useful in any provenance study, as we will see here.

Of all the tissues of the human body, few can stand the test of time: all soft tissues will decay very quickly, leaving the skeleton and the skull behind. Bones, however, are subject to at least some degree of weathering, resulting in partial isotope re-equilibrations with meteoric water and thus modification of their initial isotopic composition. We must, therefore, look to the most robust and reliable part of the body, one that remains fairly intact millennia after death: the teeth. Teeth grow early in life, and most of them are fully developed by adolescence. Some of them, however, such as wisdom teeth, will grow later and have the potential to register a different set of environmental conditions. Addi-

© Springer Nature Switzerland AG 2020
P. Alexandre, *Isotopes and the Natural Environment*, Springer Textbooks in Earth Sciences, Geography and Environment, https://doi.org/10.1007/978-3-030-33652-3_8

Fig. 8.1 Simplified map of the Stonehenge area, showing the sites where human remains were found (red stars)

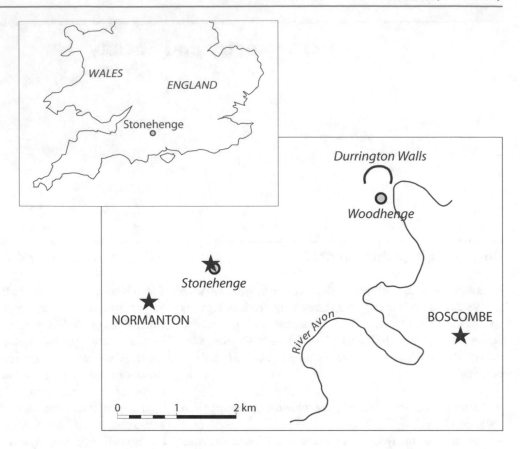

tionally, teeth are made up of two main components (Fig. 8.2): enamel, forming the extremely hard external shell of the crown of the tooth, and dentine, forming the root and the internal parts of the tooth (the pulp, containing nerves and blood vessels, will also decay very quickly). Importantly, enamel does not exchange isotopes with its environment at any point after its formation, thus recording the conditions of its formation. On the other hand, dentine does participate in isotopic exchange during life, albeit very

Fig. 8.2 Schematic cross section of a tooth, showing the three main tissues, enamel, dentine, and the pulp, which contains nerves and blood vessels and is rarely preserved

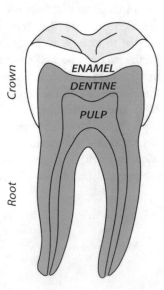

slowly. Its isotopic composition reflects the environmental conditions of the last few years. The combination of the two tissues has the capacity to decipher relatively complex histories such as the provenance of people found at Stonehenge.

Two isotopic systems were examined here: oxygen and strontium isotopes, which were both measured on the two different teeth tissues. The $\delta^{18}O$ of oxygen reflects the water isotopic composition at the time of formation of the tissue, whereas the $^{87}Sr/^{86}Sr$ ratio reflects that of the underlying rocks, soil, and vegetation used for food. This ratio is extremely sensitive to a variety of geological processes, as discussed in Chap. 4, and each particular rock formation or a region will have its own unique value. Thus, when the enamel $^{87}Sr/^{86}Sr$ ratio was measured in all individual skulls (Fig. 8.3), some had the exact same value as that for the Stonehenge area, suggesting that these people were local, whereas others had a very different and distinct $^{87}Sr/^{86}Sr$ ratio. The nearest region with a similar $^{87}Sr/^{86}Sr$ ratio is today's northern Wales, which has been interpreted as the origin for these people. This was confirmed by the $\delta^{18}O$ values that were only observed in meteoric water in the same area. Interestingly, the "Welsh" people have molar teeth enamel with a different, third, strontium isotopic composition. This suggests that during their late childhood—when their molar teeth grew—they were living in a different location, possibly in today's northern England. Finally, when the dentine $^{87}Sr/^{86}Sr$ ratio was measured, it was exactly the same as the local Stonehenge value, suggesting that these people have remained there until their death (Fig. 9.3; [3]). A third group of people was present, and their teeth enamel $^{87}Sr/^{86}Sr$ ratio and $\delta^{18}O$ values were not clearly related to a specific location: their origin is labelled as "unknown". Thus, we can conclude, with a good level of certainty, that some of the people whose skeletons were discovered at Stonehenge were local, others came from northern Wales, but spent some years in early adulthood elsewhere, and yet other people came from an unknown location.

Interestingly, and somewhat unrelated to the topic of isotopes, the actual concentration of strontium in the tooth enamel of all the people studied at Stonehenge is revealing of their diet. All of the teeth have the same concentrations (54 ± 14 ppm), which is similar to other Neolithic people. Importantly, these concentrations are also similar to modern teeth from people in the UK, but are lower than native Britons of the Roman period (>100 ppm). Strontium concentrations drop with an increasing trophic level in the diet, and the similarity with the modern diet suggests that Bronze Age people had a mixed diet, as they consumed meat and dairy products in proportions comparable to those of modern Britons. (Roman era native Britons had a diet of a much lower trophic level, as they ate many more plants than meat.)

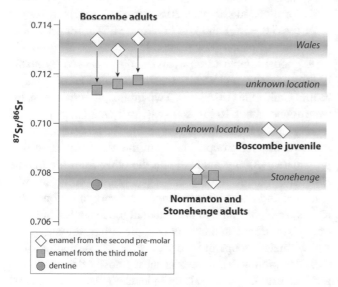

Fig. 8.3 Strontium isotopic composition from different teeth tissues for the different people whose remains were found in the Stonehenge area. The individuals from Normanton and Stonehenge were raised within the area of Stonehenge: they were non-migratory, or were at least during their childhood and early life, when their premolars and molars grew, respectively. The Boscombe juveniles originated from an unknown location, and the Boscombe adults likely originated from Northern Wales (also confirmed by oxygen isotopes). They then lived elsewhere during the growth of their molars, before coming to Stonehenge. Data from [3]

8.3 Carbon Dating

8.3.1 Methodology

Carbon has two stable isotopes, ^{12}C and ^{13}C, as we have already examined, and a curious one, ^{14}C, the hero of this story. This isotope, ^{14}C, is not stable, but has a fairly stable concentration on Earth's surface and thus retains the appearance of a stable isotope. However, ^{14}C is radioactive, which means that it disintegrates at a constant speed, to ^{14}N through a β- and a neutrino emission. Nonetheless, it is also produced at a constant rate: this production occurs in the outer reaches of the atmosphere where cosmic radiation, in the form of fast neutrons, produces ^{14}C and ^1H from ^{14}N; we shall call ^{14}C a cosmogenic isotope. (Small amounts of other light isotopes, such as ^3H, ^{10}Be, ^{26}Al, ^{32}Si, ^{36}Cl, ^{39}Ar, ^{53}Mn, ^{59}Ni, and ^{81}Kr, are produced there as well, and all of them are radioactive.)

The disintegration rate of ^{14}C is known, with a paltry 5730 years of half-life (the time it takes for half of the atoms to disintegrate to the daughter isotope). Knowing this, it is possible to calculate the age of an artefact that incorporated some amount of ^{14}C. The analytical method is relatively simple: C is converted to CO_2 and then its activity (proportional to its amount) is measured in an ionization chamber or a scintillation counter (a device consisting of secondary electron multiplier, proceeded by a special scintillation crystal converting the incoming radiation to electrons). Another, more modern, more complicated, but more precise method is the use of accelerator mass spectrometry. The sample is converted to graphite and is then accelerated, its electrons are stripped by interaction with gas, and two carbon isotopes, ^{14}C and ^{12}C, are measured by a mass spectrometer. In this case, it is the $^{14}C/^{12}C$ ratio that is used in age calculation.

Generally, the ^{14}C produced in the upper atmosphere will eventually make its way to the lower atmosphere, where it will be mostly incorporated into plants (Fig. 8.4). These plants will serve as food for people and animals, so their bones, which incorporate some carbon, will be suitable for ^{14}C dating. The plant material would itself be suitable for ^{14}C dating, as would be any other material that might have incorporated some ^{14}C. Thus, any artefact, such as a tool made of bone or wood, or the bones of people, can be dated by this method.

8.3.2 Limitations of the Method

The major limitation of ^{14}C dating is the short half-life of ^{14}C, as we had noted earlier. In practical terms, it is difficult to date any artefact older than 35,000, 40,000 years at most.

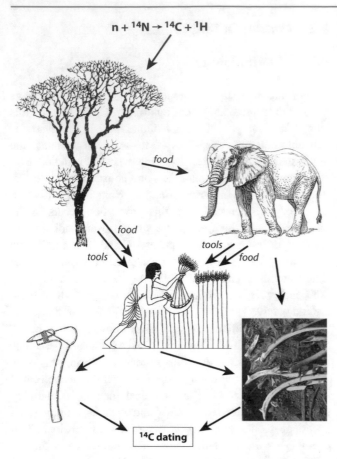

$$n + {}^{14}N \rightarrow {}^{14}C + {}^{1}H$$

food

food

tools

tools

food

¹⁴C dating

Fig. 8.4 The simplified cycle of ¹⁴C, from its production in the upper atmosphere by cosmic radiation, through its incorporation into plants and then in human and animal bones, to eventual preservation. Basically, any artefact that can incorporate some amount of carbon, even minute, can be dated by the ¹⁴C method

On the other hand, the significant increase of fossil fuel burning over the last two centuries and the atmospheric nuclear tests in the late 1950s and early 1960s have led to such dilution and production of ¹⁴C in the atmosphere, respectively, that it is impossible to date any artefact younger than 150 years.

Another limitation is the spatial and temporal variation of ¹⁴C production, due to fluctuations in cosmic rays and changes in Earth's magnetic field. Additional variations will occur, as different plants incorporate carbon isotopes at a different rate and to different extent (kinetic fractionation). Finally, contamination often leads to increased uncertainty in the amount or activity of ¹⁴C and thus of the artefact's age.

8.3.3 Example: Populating Europe

An extensive study on the settlement of Europe was conducted over the Middle East region and the whole of Europe, considering the earliest known farming communities [8].

Tools made predominantly of wood were dated by ¹⁴C dating and the results were plotted on a map (Fig. 8.5). A very clear picture emerged, in that the earliest farming communities in the study area were approximately in today's Syria and Cyprus regions, and were settled earlier than 9000 years ago. Over the next five centuries, the area of today's eastern Turkey and northern Iraq were settled. From there on in, the settlement of Europe took two paths, one due north between the Black and the Caspian seas (and thus through the Caucasus Mountains), the other northwestwardly headed through western Turkey and the Balkans. The latest areas to have farming settlements, in the British Isles and in Scandinavia, were dated at less than 6000 years old, showing that Europe's settlement occurred over more than three millennia. Most settlements likely formed following the ice sheet retreat, as the last significant glaciation period ended some 10,000 years ago.

8.4 Diet Evolution

As described in Chap. 7, diet is reflected in a person's tissues and can be assessed using stable isotopes, specifically those of nitrogen and carbon. Simplistically, the higher the $\delta^{15}N$ and $\delta^{13}C$ in a person's hair, or fingernails, or most importantly in archaeology, in bones, the higher the proportion of meat is (higher trophic levels) in their diet. Typical values for a modern person with a meat-heavy diet will be above 10‰ for $\delta^{15}N$ and above −20‰ for $\delta^{13}C$. Inversely, lower $\delta^{15}N$ (below −5‰) and $\delta^{13}C$ (below −25‰) reflect higher proportion of plants (low trophic levels).

A good example of an evolving diet assessed by stable isotopes is the prehistoric coastal population of today's central California (USA; [6]). Two groups of human remains were found about 10 km southeast of Santa Cruz, near the estuary of the Pajaro River (Fig. 8.6). These were separated into two distinct groups, one from the early Holocene (approximately 7000 years ago) and the other from the middle Holocene (about 4500 years ago). Bone collagen was extracted and analyzed for nitrogen and carbon isotopes; the results were compared to the potential food sources and their (today's) nitrogen and carbon isotopic compositions.

Two main groups of food were considered: marine (fish, shellfish, pinnipeds), terrestrial meat (mostly deer and elk), and terrestrial plants (leafy plants, nuts, and grains) (Fig. 8.7). The average isotopic composition in bone collagen of the early Holocene people was about −19‰ for $\delta^{13}C$ and about 14‰ for $\delta^{15}N$; these evolved to approximately −22‰ for $\delta^{13}C$ and about 12‰ for $\delta^{15}N$ for the middle Holocene people (Fig. 8.6). In other words, these people shifted to lower trophic level food. Using isotopic mixture calculations, similar to those described in Chap. 3, it is

Fig. 8.5 Map of Europe and the Middle East, showing the ages (in thousands of years, ky) of the earliest known farming communities. Populating the whole of Europe took more than three millennia. From [8], modified

possible to estimate that the proportion of marine food in these people's diet decreased from 77 to 54%. Their terrestrial meat consumption increased from 4 to 9%, but most significantly, their terrestrial plants intake doubled, from 19 to 38% (Fig. 8.8). In other words, the Early Holocene population relied heavily on marine foods, as they were able to procure both littoral and pelagic species. Their dependence on terrestrial food sources increased by Middle Holocene, which can be explained by climate evolution (the climate became more suitable to farming) and better technology for processing plant foods (e.g., development of mortar and pestle).

8.5   Soil Erosion

When discussing populating Europe, earlier in this chapter, we mentioned farming: indeed, farming implements in the oldest settled communities were used for ^{14}C dating. And farming has always been, still is, and will be, in the foreseeable future, the most central and essential basis of our society—even though we forget it sometimes—for a very simple reason: we all have to eat and our food comes from farming.

Now, who speaks of farming necessarily speaks of soils: simplistically speaking soil is ploughed, seeds are planted in it and crop grows and is collected. Thus, soils (along with fertilizers and water) are very important to our society and are object of intense and complex studies by countless very smart people. And many of them focus on one significant problem, namely *soil erosion*, or the progressive removal of the most nutrient-rich upper layers of soil by water, ice or snow, wind, plants and animals, and human activity. Human activity, in particular, has been particularly damaging to soil and has increased many times the global soil erosion. We are, of course, highly interested in quantifying soil erosion and isotopes—you will have guessed it—come to the rescue once again.

Help is coming from the most unlikely source: nuclear activity and specifically atmospheric nuclear bomb testing. Fallout radionuclides, resulting from the nuclear explosions, would eventually find their way down to the ground. One of

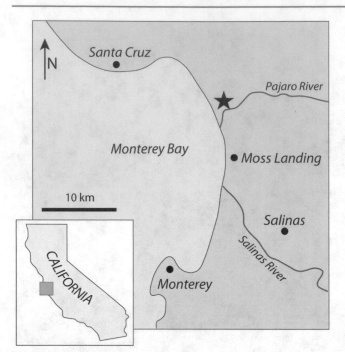

Fig. 8.6 Location of the human remains used in this study (red star; after [6], modified)

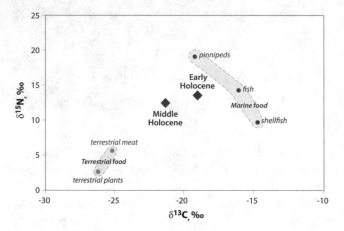

Fig. 8.7 Nitrogen and carbon isotopic composition of bone collagen for the Early and Middle Holocene, compared with isotopic compositions of the main food sources, marine (pinnipeds, fish, and shellfish) and terrestrial (meat and plants). The evolution towards more terrestrial diet is clearly visible. Data from [6]

them, ^{137}Cs, is an artificial radionuclide produced by the ^{235}U and the ^{239}Pu fission process; it has a half-life of about 30.17 years and disintegrates by β-emission. This isotope is very useful in our case, as it is very strongly fixed, or bound, to the clay minerals in the soil, so strongly indeed, that it can only be moved by physical processes, such as soil erosion [4]. Thus, the measured activity of ^{137}Cs will correspond to the movement of soil.

In practice, the activity of ^{137}Cs (in Bq/kg, reflecting its concentration) is measured in soils at different depths (at every 2 or 3 cm) down to a depth of maybe 30 or 40 cm. In non-cultivated, undisturbed soils, the activity of ^{137}Cs decreases exponentially—due to disintegration—whereas in cultivated soils it will have a homogeneous distribution with depth, down to the plough depth, after which it will decrease quickly (Fig. 8.9). If, however, the soil is ploughed and eroded, the depth profile of ^{137}Cs will be similar, but with lower overall ^{137}Cs activity (Fig. 8.9). On the basis of the reference profile, established in close proximity to the site studied, the total ^{137}Cs activity in the particular soil can be calculated; this can be compared with the total ^{137}Cs activity in the ploughed soil. The difference can be related directly to soil loss using a simple calculation: Total soil loss = ^{137}Cs loss$^{1.18}$ [7], or Total soil loss = 3.84^{137}Cs loss$^{1.55}$ [2], or any other similar empirical model.

The use of ^{137}Cs activity (but also of other fallout radionuclides such as ^{90}Sr or ^{239}Pu) as a measure of soil erosion has been in extensive use for the last four decades. The method has one limitation, though: it is fairly labour-intensive and time consuming, making it best suited for long-term monitoring programmes. More recently, another method was developed, using our good old friends, the stable isotopes of C and N [5]. The method is based on the concept that the δ^{13}C and δ^{15}N increase in depth (due to decomposition of organic matter) following very highly correlated profiles in undisturbed soils. In eroded soils, the depth profiles of δ^{13}C and δ^{15}N correlate much less: it is on the basis of this lack of correlation that the amount of soils eroded can be calculated.

Further Reading

There are a few books and many journal articles on the topic of isotopes in archaeometry; here is a brief selection:

The Archaeology of Human Bones, S. Mays, Routledge, 2010, ISBN 978-113-697177-8. The relevant part is Chap. 10, Stable isotope analysis.

Radiocarbon Dating, Second Edition: An Archaeological Perspective, R.E. Taylor and O. Bar-Yosef, Routledge, 2016, ISBN 978-131-542120-9.

Sharp, D.Z., Atudorei, V., Panarello, H.O., Fernandez, J., and Douthitt, C. (2003) Hydrogen isotope systematics of hair: archeological and forensic applications. Journal of Archaeological Science 30, 1709–1716.

White, C.D., Longstaffe, F.J., and Law, K.R. (1999) Seasonal stability and variation in diet as reflected in human mummy tissues from the Kharga Oasis and the Nile Valley.

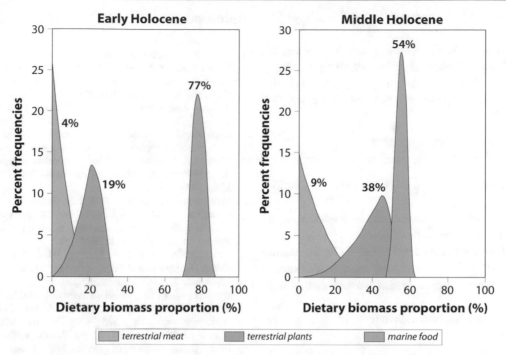

Fig. 8.8 Estimated proportions of food sources for the Early and Middle Holocene groups. The marine food sources decreased while the proportion of terrestrial plants increased significantly, probably due to the development of food processing technologies. Data from [6]

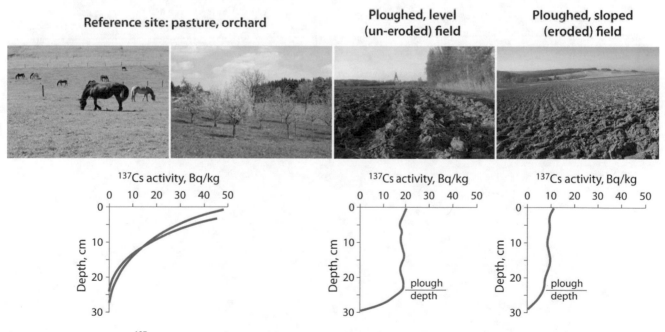

Fig. 8.9 Typical profiles of ^{137}Cs activity in soil. The undisturbed soil serves as a reference and shows an exponential decrease in ^{137}Cs activity in depth. Ploughed soils have a homogenized ^{137}Cs down to the plough depth. The measured activity can be used to calculate, with a certain level of precision, the amount of soils that has been eroded. Data from [1, 5]

Palaeogeography, Palaeoclimatology, Palaeoecology 147, 209–222.

Macko, A.A., Engel, M.H., Andrusevich, V., Lubec, G., O'Connell, T.C., and Hedges, R.E.M. (1999) Documenting the diet in ancient human populations through stable isotope analysis of hair. Philosophical transactions: Biological Sciences, 354, 65–75.

Questions

- What is archaeometry and what are its various applications?
- What material and what isotopes can be used to study an ancient population's migration?
- What is carbon dating based on? What are its limitations?
- How can we study the diet and food sources of an ancient population? What isotopes would we use and in what material?
- What methods are there to calculate soil erosion?

References

1. Andrello AC, Guimarães MF, Appoloni CR, do Nascimento Filho VF (2003) Use of Cesium-137 methodology in the evaluation of superficial erosive processes. Braz Arch Biol Technol 46:307–314
2. Campbell BL, Loughran RJ, Elliott GL, Shelly DJ (1986) Mapping drainage basin sources caesium-137. IAHS 174:437–446
3. Evans J, Chenery C, Fitzpatrick AP (2006) Bronze age childhood migration of individuals near Stonehenge, revealed by strontium and oxygen isotope tooth enamel analysis. Archaeometry 48:309–321
4. Livens FR, Loveland PJ (1988) The influence of soil properties on the environmental mobility of caesium-137 in Cumbria. Soil Use Manag 4:69–75
5. Meusburger K, Mabit L, Park J-H, Sandor T, Alewell C (2013) Combined use of stable isotopes and fallout radionuclides as soil erosion indicators in a forested mountain site, South Korea. Biogeosciences Discuss 10:2565–2589
6. Newsome SD, Phillips DL, Culleton BJ, Guilderson TP, Koch PW (2004) Dietary reconstruction of an early to middle Holocene human population from the central California coast: insights from advanced stable isotope mixing models. J Archaeol Sci 31:1101–1115
7. Ritchie JC, McHenry JR (1975) Fallout Cs-137: a tool in conservation research. J Soil Water Conserv 30:283–286
8. Silva F, Linden MV (2017) Amplitude of travelling front as inferred from ^{14}C predicts levels of genetic admixture among European early farmers. Sci Rep 7:11985

Forensics

9.1 What Is Isotopic Forensics?

We have already mentioned, in Chaps. 5 and 7, that isotopes —and in particular, stable isotopes—can be used as a forensic tool. Fundamentally, and as we have clearly stated earlier (Chap. 1), any process or reaction that has affected material will be reflected in the isotopic signature of this material. From this point of view, using isotopes (or another investigative tool) in the natural sciences equates the investigation conducted by any crime-fighting agency: natural scientists are investigators engaged in a scientific whodunit.

In forensic science, the most common way of using isotopes is as provenance tool: each and every material object will reflect the environment in which it formed or with which it interacted, and the conditions of these interactions. If we obtain the isotopic composition of an object and have an idea about the prevailing physical conditions, we can have a great level of certainty about the provenance, or source location, of that object.

Just like in any science, where there are many discrete or variously overlapping domains and disciplines, so it is in forensic science, where isotopes are applied to an amazingly wide array of activities and materials. We might be speaking of counterfeit luxury goods (a several billion dollars industry), illicit drugs tracing, food and drink authentication, origin of an unknown person, etc. From this wide array, we will present a few select examples, but let us first consider an important factor. We must first examine the discrimination between the environmental media (air, soil, water) and organic matter (plants, animals), as well as the effects of processing on isotope fractionation.

We observed in Chap. 1 that fractionation in organic material and living organisms is always kinetic, as equilibrium is not, or rarely, achieved. As a result, and simplistically, the extent of fractionation is higher and less predictable than in equilibrium fractionation, as it is affected by a range of physical and chemical factors. Often, the extent of kinetic fractionation between air or water, on one hand, and organic matter, on the other, is assessed empirically, as shown in the examples of Fig. 9.1. We rely on such correlations to assess the environmental isotopic conditions from the isotopic signature of the organic matter that was analyzed. Thus, every time we analyze the isotopic composition of a material, we have to account for the discrimination between the natural environment and the material in question, in order to obtain the isotopic signature of the environment in which this object formed or interacted with.

On the other hand, processing of a product may affect its isotopic composition: the resulting change may vary from negligible to very significant. These effects may be further complicated by processing in a different location, or country, from the one where the product originates. Sometimes the effects of processing are well understood and quantified, but sometimes it is very difficult to properly measure them.

9.2 The Origin of Bottled Water

The origin of bottled water is one of the most straightforward examples of forensic application of stable isotopes. There is no fractionation or discrimination during the bottling process, resulting in the bottled water having the exact composition of the meteoric water of the source region. As we noted in Chap. 3, hydrogen and oxygen isotopes of any meteoric water are well established and known, and it is thus fairly simple to establish the provenance of any bottled water and investigate if the origin information provided on the label corresponds to the reality.

This is a fairly simple exercise, to the point that I had my students do it as a project in this course: the data given here come from such an exercise. The students collected a variety of bottles of water from several supermarkets around town that represented a range in (perceived) quality and in price. They then analyzed the waters, obtained the hydrogen and oxygen isotopic composition for each sample, and plotted them on a δ^2H versus $\delta^{16}O$ diagram (Fig. 9.2). The results show a fair bit of variability and other similar studies have

© Springer Nature Switzerland AG 2020
P. Alexandre, *Isotopes and the Natural Environment*, Springer Textbooks in Earth Sciences, Geography and Environment, https://doi.org/10.1007/978-3-030-33652-3_9

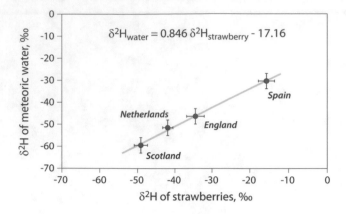

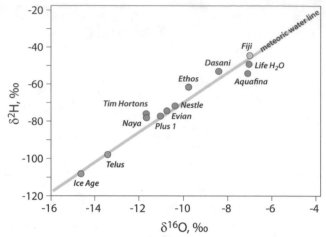

Fig. 9.2 Examples of the isotopic composition of commercially available bottled water from different producers. Two samples, with significantly lower δ^2H and $\delta^{16}O$ values (Telus and Ice Age) come from actual glaciers, which explains their isotopic composition

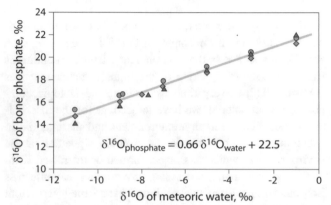

Fig. 9.1 Typical examples of empirically derived discrimination factors for strawberries and for bone phosphate: the hydrogen or oxygen isotopic composition of meteoric water can be reliably derived from that of the product. As the meteoric water will be very different between northern (Scotland) and southern (Spain) growth locations, so will be the isotopic composition of the strawberries, and any mislabeling would be easily detected. Similarly, the isotopic composition of the bone phosphate of a person will clearly reflect the location where that bone grew. Data from Meier-Augenstein [7], Daux et al. [2], Longinelli [5], and Luz and Kolodny [6]

discovered similar range of isotopic compositions. Importantly, the students compared the results with the water origin, as stated on the label of each bottled water sample, to see if the producers fairly represented the origin of the waters. In each case, the stated water origin corresponded isotopically exactly to the meteoric water in the source region, which means that no producer mislabeled their product. (The students were a little disappointed: they had hoped to catch someone cheating…)

Even the extreme values obtained by this exercise were logical. One sample (Fiji) had the highest δ^2H and $\delta^{16}O$ values; the origin of the water was stated on the label as "South Pacific". Another two samples (Telus and Ice Age) had the lowest δ^2H and $\delta^{16}O$ values and the labels claimed that the water originated from actual glaciers, one in Alaska (USA) and another in Northern British Columbia (Canada). In all three cases, the obtained isotopic compositions are

completely compatible with the meteoric water compositions in these geographic areas.

It is noticeable that several samples deviated from the global theoretical meteoric water line (Fig. 9.2). This is not surprising at all, but rather to be expected and as such illustrates what we discussed in Chap. 3: firstly, each geographic region will have its own meteoric water line, and secondly, local conditions and the effects of limited evaporation will result in these small but noticeable deviations from the theoretical meteoric water line.

Authentication of bottled water using oxygen and hydrogen isotopes is, generally speaking, relatively straightforward: there is no fractionation during the bottling process. Only one exception exists: when producing artificially carbonated water, CO_2 is added to the water during bottling. The oxygen in the water will isotopically equilibrate with that in CO_2, modifying the $\delta^{18}O$ of water. Oxygen isotopic composition of atmospheric CO_2 is always enriched in ^{18}O relative to the meteoric water, as we saw in Chap. 6, which means that carbonated water will always have a higher $\delta^{18}O$ than before carbonation.

9.3 Wine Authentication

It is now the time to complicate things a bit and move from water to wine. Again, this is a very economically significant business, with many producers from around the globe engaged in a fierce competition. Do they all use the finest grapes available in their region, or those that originate from the most reputed and valuable localities? Do the producers at least all use exactly the grapes they say that they used? Or are there perhaps some who would cheat by mixing grapes from different regions (some of those grapes would be much cheaper), without disclosing it on the label? Many countries take winemaking very seriously and mete stiff penalties to those who lie about the variety of their grapes, the wine-making processes, or the geographic origin of their wine. Unsurprisingly, these countries rely heavily on both stable and radiogenic isotopes to make sure that those cheaters are caught and punished.

Contrary to water, though, wine is a very complex product, resulting from the combination of several complicated, elaborate, and protracted processes. The factors affecting the fractionation of oxygen, hydrogen, and carbon isotopes (often the preferred elements in wine authentication) are numerous and can be summarized in two main groups: biotic and abiotic fractionation effects [1].

The main biotic effect is the mechanism of sugar photosynthesis, which we find most often in C_3 or C_4 plants, as seen in Chap. 5. Importantly, all varieties of grape are C_3 plants, whereas the main sugar-producing plants (e.g., cane and corn) are C_4 plants. Therefore, if a producer artificially introduces sugar into the wine during the winemaking process (which is mostly forbidden), this will be clearly reflected in the carbon isotope composition of the wine.

Other biotic fractionation effects, affecting the isotopes of O, H, and C, are the variety of grape used, the alcohol fermentation processes, and the addition of yeasts during winemaking.

The abiotic fractionation effects are related to the geographic location of the grape growing region, on the general climate conditions, on the specific weather conditions during the grape growing season, on the actual date of grape harvest, and on any specific treatment the wine may have undergone during its making. We have discussed in some detail how geography and weather affect the isotopic composition of meteoric water (Chap. 3), and thus the meteoric water's isotopic composition will be reflected in that of the wine, albeit modified. However, the overall implication of the abiotic fractionation effects will be that there is not one standard or stable isotopic composition for any certain wine (in particular, for hydrogen and oxygen isotopes). Rather, each season and each vintage will have different isotopic compositions (Fig. 9.3).

By now we have seen how two common frauds can be detected, that of adding sugar to the grape mush before fermentation, and that of mislabelling the vintage year. The first of the two can also be detected by using hydrogen isotopes of wine ethanol, as they will also have a distinct isotopic composition.

Other common malpractices that we will consider here are:

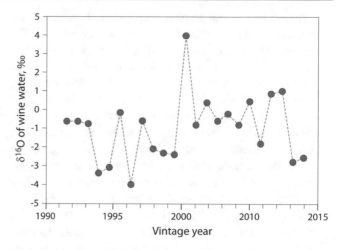

Fig. 9.3 Variations in the average $\delta^{18}O$ of wine water from Franconia, Germany, due mostly to climate and weather variations (from [1], modified). The main implication is that measuring the wine water $\delta^{18}O$ of wine from a given region is a good tool in detecting vintage fraud (claiming that the wine is from another, often superior, vintage year)

– Acidification by tartaric acid. Only tartaric acid extracted from grapes is allowed for acidification, as opposed to synthetic tartaric acid. Luckily, the oxygen and carbon isotopic compositions of natural and synthetic tartaric acids are quite distinct. People have even been able to identify the geographic region and the year of production of natural tartaric acid [4].

– Adulteration by exogenous glycerol. This used to be a common offence, as many winemakers believed (and some still do) that more glycerol makes the wine taste better. There is a certain amount of naturally produced glycerol, during fermentation, and that is all that is allowed to be present in wine: adding artificial glycerol is forbidden. Again, each time we add a foreign substance —such as when performance-enhancing drugs are used, as seen in Chap. 5—the isotopic composition of that substance will be significantly and clearly different from the isotopic composition of other substances in wine. In this specific case, we are able to detect the introduction of exogenous glycerol through the use of oxygen and carbon isotopes.

– Addition of carbon dioxide. Any CO_2 present in sparkling wine must be produced in the bottle during the second fermentation. However, many producers are tempted to add some extra CO_2, which often originates from industrial activity. This foreign CO_2 has a very distinct carbon isotopic signature ($\delta^{13}C$ above $-10‰$ or below $-29‰$, depending on the source) from a fermentation-produced CO_2 ($\delta^{13}C$ between -17 and $-26‰$), making it very easy to detect.

– Mislabelling of geographic origin. This is a very common fraud, and one that is relatively easily detected. We noted earlier that each geographic region will have specific $\delta^{18}O$ and δ^2H values for its meteoric waters. This isotopic composition will also be reflected in the wine, despite the biotic fractionation factors mentioned above (Chap. 3). If that were not sufficient, we have another isotopic tool: radiogenic isotopes. As we discussed in Chap. 4, Isotope Geochemistry, the $^{87}Sr/^{86}Sr$ ratio of each rock will be different as a result of its timing and conditions of formation and evolution. Thus, even within a relatively small region, there will be a very significant variety of soil $^{87}Sr/^{86}Sr$ ratios that reflect those of the rocks beneath. Thus, wines from any particular location will have their specific $^{87}Sr/^{86}Sr$ ratio, inherited—with some minimal fractionation—from that of the soil in which the grape plants grew.

9.4 Illicit Drugs

For some very obvious reasons, law-enforcing agencies in many countries are highly interested in knowing the provenance of illicit recreational drugs available in their countries. They want to be able to trace the drugs back to their originating point in order to more successfully combat the trade routes. And, you have guessed it, stable isotopes are an extremely powerful tool in ascertaining where the drugs were grown and produced.

Let us consider three major illicit drugs (or rather drugs with a very significant proportion of illicit use): heroin ($C_{21}H_{23}NO_5$), morphine ($C_{17}H_{19}NO_3$), and cocaine ($C_{17}H_{21}NO_4$). The first two are produced by processing opium extracted from the opium poppy, whereas cocaine is extracted from the *Erythroxylon coca* plant. In all three drugs, provenance will have a clear effect on the carbon and nitrogen isotopic composition. Carbon isotopes are affected by humidity and soil water availability, whereas nitrogen isotopes are affected by precipitation, land-use history, fertilization, and microbial N_2 conditions. On the other hand, even though hydrogen and oxygen isotopes are a very good provenance tool because of the unique meteoric water signature of each region (Chap. 3), they are less of use in this case. The processing of raw material (opium or coca leaves) often occurs in a different country from where the plant was grown, using waters with different isotopic signature. In this sense, it may be possible to detect the origin of the plants using carbon and nitrogen isotopes (Fig. 9.4), or the location of processing, using hydrogen and oxygen isotopes.

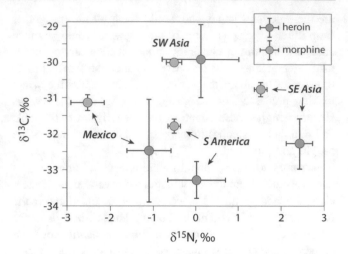

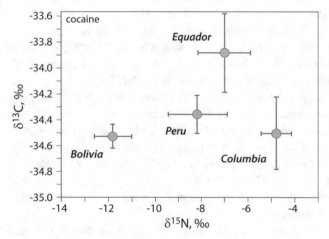

Fig. 9.4 Carbon and nitrogen isotopic compositions of three main illicit drugs, heroin and morphine, originating from the opium poppy (up), and cocaine, made from coca leaves (down), for the main growing regions. In each case, the country or region of origin is reflected in the clearly distinct isotopic composition of the drug. Re-drawn after [3]

9.5 Unknown Person

Let us imagine a situation, which can sometimes occur, that a person without any identification is apprehended by police, or that a body has been discovered and there is no way to know who the person was and where they were from. If clothing or some personal items are not available, there can be a real challenge in identifying an unknown person.

However, we have already seen, in Chaps. 7 and 8, that environmental factors are clearly reflected in a person's body tissues: this will allow us to clearly identify the region from which the unknown person comes, and even provide us with clues as to their travel history. Let us review the factors involved in identifying unknown persons using isotopic methods.

The first large category of factors is dietary differences and preferences, including the degree of processing of foods. As we observed already in Chap. 7 (e.g., Fig. 7.1), the trophic level of our main food is clearly reflected in the carbon and nitrogen isotopes of body tissues, for instance in hair or fingernails. People with preference for vegetable-only food will have lower $\delta^{15}N$ and $\delta^{13}C$ than omnivores and will have much lower δ values than exclusive carnivores (Fig. 7.1). This can significantly narrow the geographic provenance of the unknown person, based on the predominant dietary preferences in various locations, and thus would also inform us about the habits of the specific person. Further, a higher level of processing food and use of C_4 plants tends to increase the observed $\delta^{13}C$ value, which can also give hints about the origin of a person.

Another way to narrow down the geographic origin of an unknown person is to study the oxygen and hydrogen isotopic composition of hair and fingernails. As we noted in Chap. 3, each region will have a specific δ^2H and $\delta^{18}O$ of meteoric water, due to a variety of factors (latitude, distance from coast, mean annual temperature, altitude, etc.) and that will be reflected in the body tissues. Further, we can use strontium isotopes, and specifically the $^{87}Sr/^{86}Sr$ ratio (and also $^{176}Lu/^{177}Lu$ or $^{143}Nd/^{144}Nd$), to further pinpoint the location where a person has resided in the past. This ratio will be specific to any particular geographic area, as we saw in Chap. 8.

Two important factors must be considered in both diet- and location-related isotopic systems.

- Discrimination between the environment and the body tissues, as already mentioned earlier in this chapter (e.g., Fig. 9.1). Luckily, the discriminations between environment (e.g., various drink and food sources) and various body tissues are well understood and can be applied in a fairly straightforward manner.
- Time constraints. Soft tissues (e.g., muscle tissue, hair, fingernails) will reflect a timeframe of weeks to months before analyses, which informs us about the recent whereabouts of an unknown person. In contrast, bone phosphate and teeth enamel inform us about the place where the person was when these tissues grew: adult life (sometime between adolescence and years before analyses) for bone phosphate, and childhood for tooth enamel. In this way, we can have clear ideas about the locations where a person grew up, spend most or significant part of their adult life, and was in the weeks to months before analysis.

9.6 Counterfeit Goods

The general idea behind forensics of counterfeit goods is related to the other applications of isotope forensics described above. Most often, the geographic location where a particular item is actually produced (revealed by stable isotopes such as oxygen and hydrogen) is compared with where the item is supposed to have been produced. Was this luxury bag (or pair of shoes, or a bottle of perfume or of wine) made in France (or Italy, or the UK) of the finest materials or is this—often very expensive—item made elsewhere, of inferior materials, and at much lower labour and other costs? This is very easily ascertained using stable isotopes, as with many other applications.

Additionally, the processing involved in the production of any item will leave its clear in the isotopic composition of that item. As we mentioned in Chap. 1, isotopic fractionation is strongly affected by the physical and chemical conditions at which the specific process occurred. If a particular product or item was supposed to be produced at specific conditions, stable isotopes can clearly indicate if these conditions were indeed present; if not the item is likely counterfeit.

Finally, any foreign substance that is not supposed to be in a particular product will have a clearly distinct isotopic composition, as we have noted repeatedly earlier (e.g., performance-enhancing drugs, Chap. 7). We can detect a foreign substance using generally one of many stable isotopic systems. Is there some corn syrup (coming from a C_4 plant) in your maple syrup (from a C_3 plant; Chap. 5)? Are those bubbles in your sparkling wine coming uniquely from the second fermentation, or are they coming from a vulgar tank of industrial-grade CO_2?

Crucially, faking the expected isotopic composition of any product is an extremely complicated, expensive, and time-consuming task requiring a very significant amount of knowledge and specialized equipment. In practical terms, it is not possible to fake the isotopic composition of a counterfeit product to make it similar to the original one, meaning that isotopes remain the most powerful and precise way, by far, to detect counterfeit goods.

Further Reading

Food Forensics: Stable Isotopes as a Guide to Authenticity and Origin, J.F. Carter and L.A. Chesson, CRC Press, 2017, ISBN 978-149-874172-9.

Stable Isotope Forensics: An Introduction to the Forensic Application of Stable Isotope Analysis, W. Meier-Augenstein, John Wiley & Sons, 2011, ISBN 978-111-996513-8.

Questions

- What are the main applications of isotopic forensics?
- What are the consistent principles of all isotopic forensic investigations?
- What is discrimination factor and why is it important?
- How can we study the provenance of a material, and which stable and radiogenic isotopes would we use?
- How is the processing of a product affecting its isotopic composition?
- Can any faked or counterfeit product be beyond detection by isotopic methods?

References

1. Christoph N, Hermann A, Wachter H (2015) 25 Years authentication of wine with stable isotope analysis in the European Union—Review and outlook. BIO Web of Conferences 5:02020
2. Daux V, Lecuyer C, Heran MA, Amiot R, Simon L, Fourel F, Martineau F, Lynnerup N, Reychler H, Escarguel G (2008) Oxygen isotope fractionation between human phosphate and water revisited. J Hum Evol 55:1138–1147
3. Ehleringer JR, Cooper DA, Lott MJ, Cook CS (1999) Geo-location of heroin and cocaine by stable isotope ratios. Forensic Sci Int 106:27–35
4. Fauhl C, Wittkowski R, Lofthouse J, Hird S, Brereton P, Versini G, Lees M, Guillou C (2004) Gas chromatographic/mass spectrometric determination of 3-methoxy-1,2-propanediol and cyclic diglycerols, by-products of technical glycerol, in wine: Interlaboratory study. J AOAC Int 87:1179–1188
5. Longinelli A (1984) Oxygen isotopes in mammal bone phosphate: a new tool for paleohydrological and paleoclimatological research? Geochim Cosmochim Acta 48:385–390
6. Luz B, Kolodny Y (1989) Oxygen isotope variation in bone phosphate. Appl Geochem 4:317–323
7. Meier-Augenstein W (2011) stable isotope forensics: an introduction to the forensic application of stable isotope analysis. Wiley

Appendix A

See Table A.1.

Table A.1 Masses and relative abundances of the main naturally-occurring isotopes. The radioactive isotopes are given with their half-life (HL). Elements from Ununnilium (^{272}Uun) to Ununoctium (^{293}Uuo) are not listed because of their very short half-lives, between a few milliseconds and a few seconds

Atomic number	Name	Symbol	Atomic mass (AMU)	Relative abundance (%)
1	Hydrogen	^{1}H	1.007825	99.9885
		^{2}H	2.014102	0.0115
		^{3}H	3.016049	Traces (*radioactive: HL* = 12.5 y)
2	Helium	^{3}He	3.016029	0.000137
		^{4}He	4.002603	99.999863
3	Lithium	^{6}Li	6.015122	7.59
		^{7}Li	7.016004	92.41
4	Beryllium	^{9}Be	9.012182	100
5	Boron	^{10}B	10.012937	19.9
		^{11}B	11.009305	80.1
6	Carbon	^{12}C	12.000000	98.93
		^{13}C	13.003355	1.07
		^{14}C	14.003242	Traces (*radioactive: HL* = 5730 y)
7	Nitrogen	^{14}N	14.003074	99.632
		^{15}N	15.000109	0.368
8	Oxygen	^{16}O	15.994915	99.757
		^{17}O	16.999132	0.038
		^{18}O	17.999160	0.205
9	Fluorine	^{19}F	18.998403	100
10	Neon	^{20}Ne	19.992440	90.48
		^{21}Ne	20.993847	0.27
		^{22}Ne	21.991386	9.25
11	Sodium	^{23}Na	22.989770	100
12	Magnesium	^{24}Mg	23.985042	78.99
		^{25}Mg	24.985837	10.00
		^{26}Mg	25.982593	11.01
13	Aluminum	^{27}Al	26.981538	100
14	Silicon	^{28}Si	27.976927	92.2297
		^{29}Si	28.976495	4.6832
		^{30}Si	29.973770	3.0872
15	Phosphorus	^{31}P	30.973762	100
16	Sulphur	^{32}S	31.972071	94.93
		^{33}S	32.971458	0.76
		^{34}S	33.967867	4.29
		^{36}S	35.967081	0.02

(continued)

© Springer Nature Switzerland AG 2020
P. Alexandre, *Isotopes and the Natural Environment*, Springer Textbooks in Earth Sciences,
Geography and Environment, https://doi.org/10.1007/978-3-030-33652-3

Table A.1 (continued)

Atomic number	Name	Symbol	Atomic mass (AMU)	Relative abundance (%)
17	Chlorine	^{35}Cl	34.968853	75.78
		^{37}Cl	36.965903	24.22
18	Argon	^{36}Ar	35.967546	0.3365
		^{38}Ar	37.962732	0.0632
		^{40}Ar	39.962383	99.6003
19	Potassium	^{39}K	38.963707	93.2581
		^{40}K	39.963999	0.0117 (*radioactive:* $HL = 1.25 * 10^9$ y)
		^{41}K	40.961826	6.7302
20	Calcium	^{40}Ca	39.962591	96.941
		^{42}Ca	41.958618	0.647
		^{43}Ca	42.958767	0.135
		^{44}Ca	43.955481	2.086
		^{46}Ca	45.953693	0.004
		^{48}Ca	47.952534	0.187
21	Scandium	^{45}Sc	44.955910	100
22	Titanium	^{46}Ti	45.952629	8.25
		^{47}Ti	46.951764	7.44
		^{48}Ti	47.947947	73.72
		^{49}Ti	48.947871	5.41
		^{50}Ti	49.944792	5.18
23	Vanadium	^{50}V	49.947163	0.250
		^{51}V	50.943964	99.750
24	Chromium	^{50}Cr	49.946050	4.345
		^{52}Cr	51.940512	83.789
		^{53}Cr	52.940654	9.501
		^{54}Cr	53.938885	2.365
25	Manganese	^{55}Mn	54.938050	100
26	Iron	^{54}Fe	53.939615	5.845
		^{56}Fe	55.934942	91.754
		^{57}Fe	56.935399	2.119
		^{58}Fe	57.933280	0.282
27	Cobalt	^{59}Co	58.933200	100
28	Nickel	^{58}Ni	57.935348	68.0769
		^{60}Ni	59.930791	26.2231
		^{61}Ni	60.931060	1.1399
		^{62}Ni	61.928349	3.6345
		^{64}Ni	63.927970	0.9256
29	Copper	^{63}Cu	62.929601	69.17
		^{65}Cu	64.927794	30.83
30	Zinc	^{64}Zn	63.929147	48.63
		^{66}Zn	65.926037	27.90
		^{67}Zn	66.927131	4.10
		^{68}Zn	67.924848	18.75
		^{70}Zn	69.925325	0.62
31	Gallium	^{69}Ga	68.925581	60.108
		^{71}Ga	70.924705	39.892

(continued)

Table A.1 (continued)

Atomic number	Name	Symbol	Atomic mass (AMU)	Relative abundance (%)
32	Germanium	^{70}Ge	69.924250	20.84
		^{72}Ge	71.922076	27.54
		^{73}Ge	72.923459	7.73
		^{74}Ge	73.921178	36.28
		^{76}Ge	75.921403	7.61
33	Arsenic	^{75}As	74.921596	100
34	Selenium	^{74}Se	73.922477	0.89
		^{76}Se	75.919214	9.37
		^{77}Se	76.919915	7.63
		^{78}Se	77.917310	23.77
		^{80}Se	79.916522	49.61
		^{82}Se	81.916700	8.73
35	Bromine	^{79}Br	78.918338	50.69
		^{81}Br	80.916291	49.31
36	Krypton	^{78}Kr	77.920386	0.35
		^{80}Kr	79.916378	2.28
		^{82}Kr	81.913485	11.58
		^{83}Kr	82.914136	11.49
		^{84}Kr	83.911507	57.00
		^{86}Kr	85.910610	17.30
37	Rubidium	^{85}Rb	84.911789	72.17
		^{87}Rb	86.909183	27.83
38	Strontium	^{84}Sr	83.913425	0.56
		^{86}Sr	85.909262	9.86
		^{87}Sr	86.908879	7.00 (*radioactive: HL = 49.44 * 10^9 y*)
		^{88}Sr	87.905614	82.58
39	Yttrium	^{89}Y	88.905848	100
40	Zirconium	^{90}Zr	89.904704	51.45
		^{91}Zr	90.905645	11.22
		^{92}Zr	91.905040	17.15
		^{94}Zr	93.906316	17.38
		^{96}Zr	95.908276	2.80
41	Niobium	^{93}Nb	92.906378	100
42	Molybdenum	^{92}Mo	91.906810	14.84
		^{94}Mo	93.905088	9.25
		^{95}Mo	94.905841	15.92
		^{96}Mo	95.904679	16.68
		^{97}Mo	96.906021	9.55
		^{98}Mo	97.905408	24.13
		^{100}Mo	99.907477	9.63
43	Technetium	^{98}Tc	97.907216	*Radioactive: HL = 6 h*
44	Ruthenium	^{96}Ru	95.907598	5.54
		^{98}Ru	97.905287	1.87
		^{99}Ru	98.905939	12.76
		^{100}Ru	99.904220	12.60
		^{101}Ru	100.905582	17.06
		^{102}Ru	101.904350	31.55
		^{104}Ru	103.905430	18.62

(continued)

Table A.1 (continued)

Atomic number	Name	Symbol	Atomic mass (AMU)	Relative abundance (%)
45	Rhodium	^{103}Rh	102.905504	100
46	Palladium	^{102}Pd	101.905608	1.02
		^{104}Pd	103.904035	11.14
		^{105}Pd	104.905084	22.33
		^{106}Pd	105.903483	27.33
		^{108}Pd	107.903894	26.46
		^{110}Pd	109.905152	11.72
47	Silver	^{107}Ag	106.905093	51.839
		^{109}Ag	108.904756	48.161
48	Cadmium	^{106}Cd	105.906458	1.25
		^{108}Cd	107.904183	0.89
		^{110}Cd	109.903006	12.49
		^{111}Cd	110.904182	12.80
		^{112}Cd	111.902757	24.13
		^{113}Cd	112.904401	12.22
		^{114}Cd	113.903358	28.73
		^{116}Cd	115.904755	7.49
49	Indium	^{113}In	112.904061	4.29
		^{115}In	114.903878	95.71
50	Tin	^{112}Sn	111.904821	0.97
		^{114}Sn	113.902782	0.66
		^{115}Sn	114.903346	0.34
		^{116}Sn	115.901744	14.54
		^{117}Sn	116.902954	7.68
		^{118}Sn	117.901606	24.22
		^{119}Sn	118.903309	8.59
		^{120}Sn	119.902197	32.58
		^{122}Sn	121.903440	4.63
		^{124}Sn	123.905275	5.79
51	Antimony	^{121}Sb	120.903818	57.21
		^{123}Sb	122.904216	42.79
52	Tellurium	^{120}Te	119.904020	0.09
		^{122}Te	121.903047	2.55
		^{123}Te	122.904273	0.89
		^{124}Te	123.902819	4.74
		^{125}Te	124.904425	7.07
		^{126}Te	125.903306	18.84
		^{128}Te	127.904461	31.74
		^{130}Te	129.906223	34.08
53	Iodine	^{127}I	126.904468	100
54	Xenon	^{124}Xe	123.905896	0.09
		^{126}Xe	125.904269	0.09
		^{128}Xe	127.903530	1.92
		^{129}Xe	128.904779	26.44
		^{130}Xe	129.903508	4.08
		^{131}Xe	130.905082	21.18
		^{132}Xe	131.904154	26.89
		^{134}Xe	133.905395	10.44
		^{136}Xe	135.907220	8.87

(continued)

Table A.1 (continued)

Atomic number	Name	Symbol	Atomic mass (AMU)	Relative abundance (%)
55	Cesium	^{133}Cs	132.905447	100
56	Barium	^{130}Ba	129.906310	0.106
		^{132}Ba	131.905056	0.101
		^{134}Ba	133.904503	2.417
		^{135}Ba	134.905683	6.592
		^{136}Ba	135.904570	7.854
		^{137}Ba	136.905821	11.232
		^{138}Ba	137.905241	71.698
57	Lanthanum	^{138}La	137.907107	0.090
		^{139}La	138.906348	99.910
58	Cerium	^{136}Ce	135.907144	0.185
		^{138}Ce	137.905986	0.251
		^{140}Ce	139.905434	88.450
		^{142}Ce	141.909240	11.114
59	Praseodymium	^{141}Pr	140.907648	100
60	Neodymium	^{142}Nd	141.907719	27.2
		^{143}Nd	142.909810	12.2
		^{144}Nd	143.910083	23.8
		^{145}Nd	144.912569	8.3
		^{146}Nd	145.913112	17.2
		^{148}Nd	147.916889	5.7
		^{150}Nd	149.920887	5.6
61	Promethium	^{145}Pm	144.912744	*Radioactive: HL* = 17.7 y
62	Samarium	^{144}Sm	143.911995	3.07 (*radioactive: HL* = 106.0 * 10^9 y)
		^{147}Sm	146.914893	14.99
		^{148}Sm	147.914818	11.24
		^{149}Sm	148.917180	13.82
		^{150}Sm	149.917271	7.38
		^{152}Sm	151.919728	26.75
		^{154}Sm	153.922205	22.75
63	Europium	^{151}Eu	150.919846	47.81
		^{153}Eu	152.921226	52.19
64	Gadolinium	^{152}Gd	151.919788	0.20
		^{154}Gd	153.920862	2.18
		^{155}Gd	154.922619	14.80
		^{156}Gd	155.922120	20.47
		^{157}Gd	156.923957	15.65
		^{158}Gd	157.924101	24.84
		^{160}Gd	159.927051	21.86
65	Terbium	^{159}Tb	158.925343	100
66	Dysprosium	^{156}Dy	155.924278	0.06
		^{158}Dy	157.924405	0.10
		^{160}Dy	159.925194	2.34
		^{161}Dy	160.926930	18.91
		^{162}Dy	161.926795	25.51
		^{163}Dy	162.928728	24.90
		^{164}Dy	163.929171	28.18
67	Holmium	^{165}Ho	164.930319	100

(continued)

Table A.1 (continued)

Atomic number	Name	Symbol	Atomic mass (AMU)	Relative abundance (%)
68	Erbium	^{162}Er	161.928775	0.14
		^{164}Er	163.929197	1.61
		^{166}Er	165.930290	33.61
		^{167}Er	166.932045	22.93
		^{168}Er	167.932368	26.78
		^{170}Er	169.935460	14.93
69	Thulium	^{169}Tm	168.934211	100
70	Ytterbium	^{168}Yb	167.933894	0.13
		^{170}Yb	169.934759	3.04
		^{171}Yb	170.936322	14.28
		^{172}Yb	171.936378	21.83
		^{173}Yb	172.938207	16.13
		^{174}Yb	173.938858	31.83
		^{176}Yb	175.942568	12.76
71	Lutetium	^{175}Lu	174.940768	97.41
		^{176}Lu	175.942682	2.59 (*radioactive*: $HL = 37.1 * 10^9$ y)
72	Hafnium	^{174}Hf	173.940040	0.16
		^{176}Hf	175.941402	5.26
		^{177}Hf	176.943220	18.60
		^{178}Hf	177.943698	27.28
		^{179}Hf	178.945815	13.62
		^{180}Hf	179.946549	35.08
73	Tantalum	^{180}Ta	179.947466	0.012
		^{181}Ta	180.947996	99.988
74	Tungsten	^{180}W	179.946706	0.12
		^{182}W	181.948206	26.50
		^{183}W	182.950224	14.31
		^{184}W	183.950933	30.64
		^{186}W	185.954362	28.43
75	Rhenium	^{185}Re	184.952956	37.40
		^{185}Re	185.9549861	Traces (*radioactive*: $HL = 41.6 * 10^9$ y)
		^{187}Re	186.955751	62.60
76	Osmium	^{184}Os	183.952491	0.02
		^{186}Os	185.953838	1.59
		^{187}Os	186.955748	1.96
		^{188}Os	187.955836	13.24
		^{189}Os	188.958145	16.15
		^{190}Os	189.958445	26.26
		^{192}Os	191.961479	40.78
77	Iridium	^{191}Ir	190.960591	37.3
		^{193}Ir	192.962924	62.7
78	Platinum	^{190}Pt	189.959930	0.014 (*radioactive*: $HL = 469.3 * 10^9$ y)
		^{192}Pt	191.961035	0.782
		^{194}Pt	193.962664	32.967
		^{195}Pt	194.964774	33.832
		^{196}Pt	195.964935	25.242
		^{198}Pt	197.967876	7.163

(continued)

Table A.1 (continued)

Atomic number	Name	Symbol	Atomic mass (AMU)	Relative abundance (%)
79	Gold	^{197}Au	196.966552	100
80	Mercury	^{196}Hg	195.965815	0.15
		^{198}Hg	197.966752	9.97
		^{199}Hg	198.968262	16.87
		^{200}Hg	199.968309	23.10
		^{201}Hg	200.970285	13.18
		^{202}Hg	201.970626	29.86
		^{204}Hg	203.973476	6.87
81	Thallium	^{203}Tl	202.972329	29.524
		^{205}Tl	204.974412	70.476
82	Lead	^{204}Pb	203.973029	1.4
		^{206}Pb	205.974449	24.1
		^{207}Pb	206.975881	22.1
		^{208}Pb	207.976636	52.4
83	Bismuth	^{209}Bi	208.980383	100
84	Polonium	^{209}Po	208.982416	*Radioactive: HL* = 138 d
85	Astatine	^{210}At	209.987131	*Radioactive: HL* = 8.1 h
86	Radon	^{222}Rn	222.017570	*Radioactive: HL* = 3.8 d
87	Francium	^{223}Fr	223.019731	*Radioactive: HL* = 22 m
88	Radium	^{226}Ra	226.025403	*Radioactive: HL* = 1600 y
89	Actinium	^{227}Ac	227.027747	*Radioactive: HL* = 21.772 y
90	Thorium	^{230}Th		*Radioactive: HL* = 75.69 * 10^9 y
		^{232}Th	232.038050	*Radioactive: HL* = 14.01 * 10^9 y
91	Protactinium	^{231}Pa	231.035879	*Radioactive: HL* = 32.76 * 10^9 y
92	Uranium	^{234}U	234.040946	0.0055 (*radioactive: HL* = 245.25 * 10^3 y)
		^{235}U	235.043923	0.7200 (*radioactive: HL* = 703.8 * 10^6 y)
		^{238}U	238.050783	99.2745 (*radioactive: HL* = 4.468 * 10^9 y)
93	Neptunium	^{237}Np	237.048167	*Radioactive: HL* = 154 * 10^3 y
94	Plutonium	^{244}Pu	244.064198	*Radioactive: HL* = 24.1 * 10^3 y
95	Americium	^{243}Am	243.061373	*Radioactive: HL* = 432.2 y
96	Curium	^{247}Cm	247.070347	*Radioactive: HL* = 163 d
97	Berkelium	^{247}Bk	247.070299	*Radioactive: HL* = 1380 y
98	Californium	^{251}Cf	251.079580	*Radioactive: HL* = 2.6 y
99	Einsteinium	^{252}Es	252.082972	*Radioactive: HL* = 20.47 d
100	Fermium	^{257}Fm	257.095099	*Radioactive: HL* = 100.5 d
101	Mendelevium	^{258}Md	258.098425	*Radioactive: HL* = 51.5 d
102	Nobelium	^{259}No	259.101024	*Radioactive: HL* = 3.1 m
103	Lawrencium	^{262}Lr	262.109692	*Radioactive: HL* = 2.7 m
104	Rutherfordium	^{263}Rf	263.118313	*Radioactive: HL* = 1.3 h
105	Dubnium	^{262}Db	262.011437	*Radioactive: HL* = 32 h
106	Seaborgium	^{266}Sg	266.012238	*Radioactive: HL* = 2.4 m
107	Bohrium	^{264}Bh	264.012496	*Radioactive: HL* = 1 m
108	Hassium	^{269}Hs	269.001341	*Radioactive: HL* = 22 s
109	Meitnerium	^{268}Mt	268.001388	*Radioactive: HL* = 7.6 s

Data from Audi and Wapstra [1], Audi and Wapstra [2], and Rosman and Taylor [3]

References

1. Audi G, Wapstra AH (1993) The 1993 atomic mass evaluation (I) atomic mass table. Nucl Phys, Sect A 565(1):1–65
2. Audi G, Wapstra AH (1995) The 1995 update to the atomic mass evaluation. Nucl Phys, Sect A 595(4):409–480
3. Rosman KJR, Taylor PDP (1999) Isotopic compositions of the elements 1997, American Institute of Physics and America Chemical Society. J Phys Chem Ref Data 27:1275–1287

Index

© Springer Nature Switzerland AG 2020
P. Alexandre, *Isotopes and the Natural Environment*, Springer Textbooks in Earth Sciences,
Geography and Environment, https://doi.org/10.1007/978-3-030-33652-3

Printed in the United States
By Bookmasters